TIME-SCALES AND ENVIRONMENTAL CHANGE

Time is an often unstated but ever present element in all debates about environmental change; but the time-scales that the participants in these debates have in mind are often very different. For a politician long-term seldom means more than five years; a geomorphologist, on the other hand, will think of long-term in the context of millions of years. A difference in the conception of time is the subtext of many disagreements about environmental change – yet there is little debate about what terms such as long-term might mean.

This book brings together the work of a wide range of experts, working in very diverse disciplines, in order to explore the issue of time-scales and environmental change. It covers a wide range of both temporal and geographical scales; from global climatic change over many thousands of years to changes in local South African vegetation over a few decades. Each chapter illustrates the importance of explicitly addressing the issue of time when considering the environment and the importance of both considering past change before making assumptions about current trends, and of exploring fully the limitations of exercises to predict future change. The various chapters are linked together by editorial commentaries designed to highlight areas of agreement or contradiction between the contributors.

By explicitly addressing many of the assumptions about time implicit in discussion of environmental change, this book sheds new light on oft studied issues. The book does not seek to provide any clear answers about the 'right' time-scale to bear in mind when thinking about the environment but, rather, to encourage students, researchers and indeed anybody interested in environmental issues, from whatever perspective, to think about change along new and different time-scales.

Thackwray S. Driver is Teaching Research Fellow in the Geography Department at the School of Oriental and African Studies; **Graham P. Chapman** is Professor of Geography, University of Lancaster.

TIME-SCALES AND ENVIRONMENTAL CHANGE

Edited by
Thackwray S. Driver
and
Graham P. Chapman

London and New York

First published 1996
by Routledge
11 New Fetter Lane, London EC4P 4EE

Transferred to Digital Printing 2003

Simultaneously published in the USA and Canada
by Routledge
29 West 35th Street, New York, NY 10001

© 1996 Thackwray S. Driver and Graham P. Chapman

Typeset in Garamond by Solidus (Bristol) Limited

British Library Cataloguing in Publication Data
A catalogue record for this book is available from the British Library

Library of Congress Cataloging in Publication Data
A catalogue record for this book has been requested

ISBN 0-415-13252-5
ISBN 0-415-13253-3 (pbk)

CONTENTS

CONTENTS

FIGURES

TABLES

CONTRIBUTORS

William Beinart, Department of History, University of Bristol, UK

Graham P. Chapman, Department of Geography, Lancaster University, UK

Nicholas J. Clifford, Jackson Environment Institute and Department of Geography, University College, London, UK

Thackwray S. Driver, Department of Geography, School of Oriental and African Studies, London, UK

Malte Faber, Alfred Weber Institut, University of Heidelberg, Germany

James Fairhead, Department of Anthropology, School of Oriental and African Studies, London, UK

John Gordon, Hornsey, London, UK

Jean Grove, Girton College, Cambridge, UK

Melissa Leach, Institute of Development Studies, University of Sussex, UK

John McClatchey, School of Environmental Science, Nene College, Northampton, UK

John L.R. Proops, Department of Economics, University of Keele, UK

Neil Roberts, Department of Geography, Loughborough University, UK

Ros Taplin, Climatic Impacts Centre, Macquarie University, Sydney, New South Wales, Australia

Max Wallis, School of Mathematics, University of Wales at Cardiff, UK

PREFACE

Sometime during the spring 1993 the two of us had a discussion about our respective research and found some unexpected similarities in what we were at that time involved in. This book is in a way the end product of that discussion. At that time one of us (G.P.C.) had just finished writing a paper on environmental myths and international politics in the Bengal delta and the other (T.S.D.) one on policies designed to prevent over-grazing in the mountain areas of Lesotho. What we discovered was that in both cases much confusion and 'bad policy' could be directly traced back to the protagonists' unwillingness to look backwards before making a bold step forwards. Both papers also highlighted the problem of confusing short-term fluctuations with longer-term environmental change. On the pastures of Lesotho the impact of periodic drought is frequently referred to as range degradation by both government and international development agencies, whereas in the Bengal delta the finger of blame for flooding is pointed not at the monsoon but at Himalayan farmers.

The distances we respectively advocated that the protagonists needed to look back in the two cases were, however, very different. Over the issue of the international environmental politics of the Bengal delta G.P.C. argued that what was needed was a long-term vision of South Asian geological history – an understanding that the high erosion rates in the Himalayas, the high siltation levels in the Ganges and Brahmaputra, and the frequent flooding of the Bengal delta had very little to do with human agency and a lot to do with the physical geography of the region (see Chapman, Chapter 11). In the case of anti-erosion policy in Lesotho, on the other hand, T.S.D. argued that what was needed was a recognition of much more recent history – specifically the fact that current policies to prevent the perceived threat of over-grazing are almost identical to ones attempted and abandoned in abject failure in the 1940s and 1950s.

We decided that what both of us needed was some sort of framework to help us consider different time-scales when examining environmental change. The idea for the conference and then this book was essentially born out of our own need to develop this framework. Part of the problem, we decided, was that different disciplines had very different time-scales in mind when considering the environment – and, therefore, a good starting point would be to get people from different disciplines together to show how their discipline understood the time-scales involved in environmental change.

Graham P. Chapman and Thackwray S. Driver

1

TIME, MANKIND AND THE EARTH

Graham P. Chapman and Thackwray S. Driver

INTRODUCTION

This book is set firmly in the present. It is set in the present context of the debates about environmental change, and specifically the concerns over potentially damaging anthropogenic environmental change – be it at the larger scale of issues like global warming, or the smaller scales concerned with soil erosion in limited areas. But it is our conviction that many of these debates demonstrate that the protagonists have not been well informed about the differing time-scales of the phenomena examined, and that consequently both unnecessary alarm and unnecessary complacency may result. On occasion misunderstanding may even lead to pointless international political friction. The present intention of the book is therefore to explore many different aspects of time, and the manner and treatment of time by differing disciplines.

In the introduction to this chapter we start with an example of the significance of an understanding of time-scales with reference to environmental change and stability in the British Isles. There are conservationists who see the protection of particular plant communities on stable sand-dunes as a priority, as though there is or ought to be a natural permanence to what is at other scales transient and ephemeral. There are others who believe that there is a natural flora in the UK, that can admit of no new members. But the natural species diversity of the British Isles is lower than that of the nearest similar European environments, and in turn the natural species diversity of those is less than that of roughly similar environments at similar latitudes at the other side of the Eurasian landmass. A plausible explanation of this is to do with the cyclical impacts of ice advances and retreats during the Ice Age, the configuration of the basic physical geography of these areas, and the stage we have reached during the current interglacial period.

The period of the most recent Ice Age (there have been other, earlier ones in earth history) is known as the Pleistocene, which roughly speaking is synonymous with the Quaternary era of geology, and covers the last 1.5–2.0 million years of earth history. During this time the ice has advanced and

1

retreated several times; globally there are indications of about seventeen major cold periods, many of which may have supported valley glaciers and ice-sheets in Britain. The actual number of times is uncertain because successive advances erase most of the evidence of previous ones. The duration of the cold periods has been about 100,000 years, with interglacial warm periods in between when conditions in Europe have been as mild as now or milder. These periods have lasted approximately 60,000 years.

Current theory suggests that as the ice advances, some species migrate towards the tropics, sometimes in new combinations – although the tropical rain forests survive in refugia. As the ice retreats, the belts readvance. In southern Scandinavia the birch and the pine reappear 10,000 years ago, the oak forest 7,000 years ago, and the beech trees just 2,000 years ago. How does this process affect Western Europe and Britain? Western Europe is a peninsula cut off on its southern flank from Africa by the Mediterranean Sea. It is also a landmass marked by extensive east–west mountain chains, such as the Pyrenees and the Alps. As the ice advances from Scandinavia and highland Britain, so simultaneously it advances from the Alps and the Pyrenees. Thus, cut-off, most of the southwards-moving ecosystems have nowhere to retreat to, and even those that move south down the Rhone valley face the Mediterranean Sea. Successive ice advances and retreats might therefore eliminate many pre-Pleistocene species, and recolonization has to take place from more distant refugia such as the Balkans.

At the height of the last glaciation sea levels were 100–150 metres below current levels. As the ice retreated, for a time Britain was still connected to the continent of Europe, but as sea levels recovered, the land bridge was breached about 6,000 years ago, and Britain became an island.

To put the case simply, because of its peculiar configuration and its latitude, stable conditions have to prevail for a very long time in Europe for species diversity to be augmented. There simply has not been enough *time* for diversity to be re-established, and in Britain this situation has been aggravated by the presence of the sea barrier.

Since the ice sheets covering Britain retreated perhaps 15,000 years ago, we could perhaps imagine that we are in only the very first stages of a new interglacial thermal period, and that at some stage in the future the ice will return, again condemning 30 per cent of the earth's land surface to the practically lifeless sterility of current Antarctica, and yet again gouging soils and exposing bare rock. Mankind has been on the planet throughout the Pleistocene, mostly deriving an existence from hunting and gathering. It is only in the current post-glacial phase that mankind has developed settled agriculture, urban settlements, writing, and, in the last two hundred years, an industrial economy. This most recent and brief period of earth history therefore encompasses the ancient civilisations of Egypt, Mesopotamia, the Indus Valley, and China, the subjects of intensive archaeological study as we try to recover and understand our past. To someone focusing on this archaeological record, the founding of the First Dynasty of Egypt in about 3100 BC apparently happened a long time ago, almost unimaginably long ago, because a human being finds it difficult to imagine the experience of feeling the passing of 5,000 years.

By extension, the retreat of the ice from Europe 15,000 years ago seems so long ago that it is almost before time began. But, if the reference point is not man looking at the earth, but the earth itself, then 15,000 years takes on a very different feeling. If we recalibrate the earth's actual age of 4.6 billion years to be represented by a whole year, then the last 15,000 years become less than the last two minutes of that year. Seeing a high-speed film of the earth's history that started on 1 January at 00.00 hours, the ice retreats from Europe on the following 31 December at 23.58 hours. Suppose we have completed watching this film for a whole year and are now at midnight on 31 December and are preparing to drink the champagne poured in the last two minutes; these last two minutes seem like part of the present.

Students of geology and geography learn to develop intellectually a long time perspective. The majority of people are not exposed to such a perspective, and develop their understanding of time within very different frameworks. The most basic and the one which a child develops first, from its prenatal days, is the circadian rhythm, determined not only by its own biological clock, but by the parental rhythms surrounding the child. As we grow older, we learn other perspectives and understanding, as birthdays come and go, and as we query the chain of human being through knowledge of our parents, and their parents.... Many of us will extend our understanding from that to the collective memory of the culture to which we belong. As Cicero observed of the necessity of understanding history, 'Not to know what took place before you were born is to remain for ever a child.' To understand what happened before you were born requires an imaginative leap to extend an understanding of time beyond your own lifespan. But we must remember that this is still the time-scale of human history, to which Cicero refers, not the time-scale of the planet earth. That time-scale is so different it is almost unimaginable: the earth has existed for 4.6 billion years already, and may easily exist for another 4.6 billion and more. We can hope that the biosphere will continue to live on earth for that time too – and even if some intelligent human descendants of some form survive, the one thing we can be sure of is that they will be a distinct species that has evolved new abilities.

We hope enough has been said to persuade the reader that a closer examination of the concepts of time is worthwhile, for that is where we go next.

THE HUMAN IDEA OF TIME

We talk blithely of living in four dimensions: the three dimensions of space, and the one dimension of time. Space is something we can apprehend directly

– we can see near objects, set against further horizons. We can reach out with our hands and feel the width of a tree trunk, or try to put our hands round its girth. We can also standardise our measurements internationally, against the wavelength of light. In this we are using the sense organ with which evolution has equipped us best to negotiate the spaces around us – the eye. But there is no sense organ associated with time. Apparently we experience the passing of time, and we can compare our experiences. Older people typically feel that 'time is passing faster', perhaps because each successive year is a smaller part of their accumulating length of life. But this has not said what it is we experience, and how we sense it. Time is colourless, odourless, invisible, and silent. So what is it?

Time and intuitive human experience

The identification of time with subjectivity is probably as old as philosophy: all that we can touch or handle ... has shape or magnitude, whereas our thoughts and emotions have duration and quality, a thought recurs or is habitual, a lecture or a musical composition is measured upon the clock.

(Yeats 1937, p. 71)

For the moment it will suffice to distinguish between an absolute view of time and a relative view. The absolute view of time is of time as the fourth dimension, an axis marked out with intervals, and along which events can be located. The other view is that time is relative – relative to changes that mark its passing. For most of human history, it is relative time that has mattered. Relative time can be much more complicated than absolute time: since it is not constrained to a single dimensional axis, repetition is possible. Cyclical time – the repeating of a day, or a week, or a year – is part of such a time frame. In absolute time 16 December 1994 cannot be repeated. But in relative time, Fridays keep returning.

In relative time, duration is measured in relation to something. For most of human history these measures have been quite crude. A society that lives by daylight knows enough about the passage of the day to indicate when to go to or come from the fields. If for other purposes a measure is needed, some common activity will often suffice. In India a common rural unit of time was the time taken to boil a pot of rice. In ancient Egypt time was measured by shadow clocks – a kind of early sundial – or by water dripping from a tank, or by sand flowing in the hour glass. The length of time measured of course depended upon an arbitrary gradation or calibration of the contrived event. All these times stress the human dimension and human utility. Perhaps the most significant duration of all is the very idea of a human lifetime or of human life span. The duration of a human being's life is in a direct sense the maximum length of time which he or she can

4

directly experience: for the sake of the argument, let us say the biblical three score years and ten.

If this is the human scale of time, then quite clearly the larger questions about time and change will be understood from this perspective. The passing of history is recorded in the Old Testament in terms of the generations that are begotten of each other since Adam. And to know where we are in time at present, the Christian West uses a calendar which counts in years the age (at the time of publication of this book this is 1,996 years old) of the one human being who has reputedly escaped death and a normal seventy-year lifespan. This quite clearly establishes the very idea of 2,000 years as in some sense being 'a lot', or 'a long time', since obviously it is now 1,996 years since 'proper' time began.

If in human terms 2,000 years may be long, what is short at the other end of the human scale? It seems that there is a short duration below which a human being cannot distinguish. Psychologists testing people's ability to estimate the length of short intervals conclude that this is about 0.1 seconds. This, it would seem, is tied up with limits in our ability to process information.

The question then arises, how do we even recognise that there are shorter and longer time spans than these scales suggest? A separate question follows which is just as important: how do we experience or imagine these shorter and longer time intervals? Quite clearly we have to use methods of indirect observation in the first instance, and in the second instance we have to be able to imagine things beyond common-sense experience. For many phenomena we use indirect methods and then recalibrate them to our own time-scales. Some we can apprehend through devices such as high-speed photography and slow play-back. Conversely, we also need very slow-lapse camera work and subsequent speeding up to appreciate relatively slow phenomena, such as the growth of a plant. In other words, we calibrate these other processes to the human dimensions of time experience. Unfortunately we do not have a time lapse sequence of photographs of the last Ice Age advancing and retreating over Europe. By contrast, the experience of large spaces does not present us with quite the same difficulties. Not only can we see large distances from suitable vantage points, we can also traverse by foot, by car, or by aeroplane vast distances which allow us to sense the significance of large spaces.

Tolstoy observed that to accept the Copernican theory that the earth revolves round the sun, we have to suspend our obvious sensation that the sun is going round the earth. Our obvious temporal sensations of the earth are that continents do not move, that mountains do not change, that 'ol' Man River, he just keeps rolling along'. These are the assumptions that underlie most traditional accounts of the earth – thereby, in a tautological way, reinforcing in folk memory the human scale of earth history. Next we consider two of these cosmologies, the Judaic–Christian and the Hindu, which still underpin common understandings of the earth by many people in the West and in the East. The scientific refutation of these accounts has occurred considerably more recently than the refutation of either flat-earth theory or the pre-Copernican solar system.

The cosmologies of mankind

As far as we know, mankind is the only species on earth to have achieved consciousness – defined as self-consciousness. Moreover, the human species is the only one which has evolved a language sophisticated enough for the development of abstract concepts, and for their verbal communication to other members of the same species. It seems probable these two properties – of self-awareness and communication – are evolutionarily interlinked, and that they contain within themselves the need and the means to explain the origins of the world in which we find ourselves, the origins of man, and the relationship between the two. The most ancient stories of the most ancient societies are stories of the creation. By definition they have to be stories about space – the creation of the world as matter – and about time, since they are narratives about events ordered in time. Most of these stories are in a modern scientific sense objectively wrong – though nearly all contain some surprising aspects of 'the truth', which hint at a deeper understanding of the world than is apparent at first sight. The impact of deep and old folk histories itself runs deep and long. The extent to which our values and attitudes may be guided by an unacknowledged, older, and wrong 'truth' rather than a modern scientific new 'truth' is clearly apparent only when we start looking at the larger patterns of human behaviour. There is considerable evidence in this book that outdated cosmologies can still sway the professional as well as the popular mind.

Here we will for a moment reflect on two rather different ancient cosmologies: the Judaic–Christian understanding provided by Genesis and the subsequent events of the Old and New Testaments, and the under-standing provided by Hinduism. Judaism is a dogmatic religion, in the sense that it has a revealed source of truth – the Old Testament – and that it postulates a God who exists outside of and before the Universe. This God created the world in six days, and on the seventh he rested. The oceans are created first, the heavens (the firmament) next. On the third day the land is separated from the waters and vegetation is created on the land. On the fourth day the sun, moon, and stars are created to mark the passing of day and night. These are followed on the fifth day by the creatures in the waters and the fowl of the air. On the sixth day he created the beasts of the field and finally, as the last act, he placed man in the garden of Eden: 'And God said, let us make man in our image, after our likeness: and let them have dominion over the fish of the seas, and over the fowl of the air, and over the cattle, and over all the earth, and over every creeping thing that creepeth upon the earth.' This Old Testament order of the creation is not entirely wrong. The seas predate life, the fish and the fowl predate the cattle, and in the last act comes man. But it seems odd to place the vegetation of the land before the creation of the sun; and whales are made synonymous and contemporary with fish.

The story has many pointers in it.

1   Although a contrary reading is possible, it seems that the concept of a day, or more correctly what constitutes a day's length, pre-exists the creation of the world – hence God managed to create it in seven days. This idea of time as a pre-existing axis, a Newtonian coordinate into which things are mapped, is a view of absolute time that is probably the easiest to comprehend and the one most instinctively articulated in modern society – dominated by clocks – though it may not be the most intuitive way of experiencing time.

2   A beginning is located in this absolute time. Beginnings and ends are something which concern all human beings. Stories have beginnings and ends – and to some extent this book does. The concept of a beginning and an end is obviously human – as any abstract concept is – but presumably explicable because of our own apprehension of our own birth and death. Accordingly, in the Christian West we have birthdays – days located in the axis of absolute time. The earth's birthday implied by Genesis should have a date too. Archbishop James Ussher calculated in 1654, and his calculations were confirmed by the vice-chancellor of Cambridge University, that 'Heaven and Earth centre and circumstance were made in the same instance of time, and clouds full of water and man were created by the Trinity on the 26th of October, 4004 BC at 9 a.m. in the morning' (quoted in Chorley et al. 1964, p. 13).

3   The manner of the end of the earth is conjectured in both the Old Testament and the New, and both in a sense speak of a judgement day, when God will bring his wrath upon the earth, and the value of people's lives will be judged. The end will be caused by an external event – God. It is not inherent in the earth itself, since the earth itself is inherently unchanging. The date of the ending is not fixed precisely, but is in some imaginable human future. Down through the ages there have been many people who have tried to be more precise, who have prophesied the end within their own lifetime and the lifetime of others living. Whichever way it is looked at, the end has to come within a time-scale which relates to human lifespans and comprehension – to be neat, perhaps something like on 31 December, AD 3000 at 9 p.m. Many modern environmentalists also foresee an end to human life as we know it on such a human time-scale.

Hindu cosmology could hardly be more different. The idea of cycles and repetition dominates. There are many universes, new ones forming out of the collapse of older ones. They go through phases in which creation exceeds destruction, as order emerges out of chaos, to be followed by the descent again into chaos. Of the many divisions and subdivisions of time, one, the Kalpa, is most important. It represents one day in the life of Brahma, the life force of the universe. It is equivalent to 4,320 million years – close to the age of the earth.

Hinduism has many gods representing many faces of the universe. The god Vishnu represents the Creator, and the god Shiva the Destroyer. Both are necessary, since the cycle of creation and destruction is unending – but at different times different aspects are dominant. The Kalpa is subdivided into Maha Yugas, each of these being subdivided into a further four Yugas, of which the last stage is a Kali, or Black, Yuga in which destruction is uppermost and the descent into chaos occurs – and according to sages we are in one now. That this is so can be confirmed – earthquakes strike, bridges collapse, droughts are followed by floods, dams collapse, the plague strikes – and that is the way of things.

The rhythm of creation and destruction is the sleeping and waking of Brahma. Brahma is not separate from the universe, is not outside it. (S)He is synonymous with it – and is represented in purest form by pure energy. All of creation is part of this, and all, including each animal, plant, rock, and person, is but a fragment of the whole. Man is not the last part of creation, merely a simultaneous fragment of the whole. Each human body is related to an *atman* (a 'soul') which has no real beginning, and which progresses through a cycle of births and rebirths – reincarnation – not moving towards a real end so much as blending with Brahma in a higher state.

Such a cosmology looks at time as a relative concept, marked by the changes of the universe that occur, but also looks at it as a cyclical concept, marked by daily cycles, monthly cycles, annual cycles, cycles of lifetimes, and cycles of universes.

TIME AND THE NATURAL WORLD

In the post-Renaissance period the Baconian idea of science led rapidly to a distinction between man and the natural world, and to the idea that man could be a separate external observer of that world, and that nature was 'other', there to be discovered. In this section we sketch how an understanding of age and time has emerged in the natural sciences.

The discovery of the age of the earth

The account of the origin of the earth as given in Genesis was largely uncontested in Europe until two hundred years ago. Astute thinkers did, however, wonder about some aspects of the account. Observers of the natural world were worried by odd observations in nature – particularly the occurrence of sea shells in large quantities in soft rocks high in the Alps. Genesis does provide a sort of answer to the problem – in that the flood which Noah survived in his Ark quite clearly reached to the tops of the mountains – but it lasted only forty days in the biblical account. To a believer, the shells were thus easily explained as clear proof of God's account in Genesis. But others, such as Leonardo da Vinci, were not so sure. He worked out how fast the clams and other invertebrates could propel themselves, and came to the conclusion that in forty days they would hardly have got more than a few hundred yards from the normal coastline. He also observed that they occurred in profusion which suggested they had bred where they lay – and again, breeding takes more than forty days.

In the mind of the acute observer, this could also raise questions about the age and formation of mountains. In general these were held to have been created in order that they could provide valleys between them for rivers to flow in. Some people postulated that the mountain valleys had been created by currents in a great sea that had covered the earth – and indeed of course this fitted Genesis and the account of Noah.

In the eighteenth century, as the industrial revolution took off, so the sophistication and scale of mining rapidly increased. Prospectors needed people who recognised rocks, and who could predict from sequences of rocks. The same was true for the new skills of canal building. It was important to know whether water might leak, and where it might be easily contained; which areas would have reliable water supplies through the summer, and which would dry out. The first survey geological cross-sections in the UK were drawn up by a canal builder, William Smith (1769–1839). All of this led to an increase in practical knowledge of geology, and a primitive recognition of a rather general sequence of rocks.

In Germany Abraham Werner (1749–1817), a mine engineer, became professor and founder of a new science which he termed 'geognosy' – literally 'earth-knowledge'. Werner noted particularly that there was usually an order to the sequence of rock types, and that this order was vertically expressed. Base rocks were hard and crystalline, and underlay transitional rocks such as slate (which contained the first organic remains). These in turn underlay the 'flötz' rocks, such as limestone, chalk, and basalt (not recognised as igneous intrusive). Finally there were derivative rocks like sand and clay. Werner theorised that all of this had come about through a sequence of precipitation from a primordial sea. At the last stage, there had been earthquakes, cracks had appeared, and much of the sea had gone down into the earth's interior. The Werner camp became known collectively as the Neptunists.

This seems to fit Genesis – and one of the great triumphs of his theories was that they never confronted the theologians too head-on. But people pointed out discrepancies: the fish were supposed to have been created after the seas – so why were there organic remains in the lower rocks; and how did volcanoes fit in this scheme of things? Werner also realised that there were some discontinuities – termed nonconformities by modern geologists –

where rock strata which have been angled upwards are planed off, and new strata are laid across this surface. He thought they were local accidents caused during the time when the earth trembled and the sea went down its cracks – and for the same reason not all rocks were horizontal. The significant points about his theories include the fact that they give the earth one start, one 'condensation', and one final form – which is essentially unchanging. The difficult parts like volcanoes were explained as subterranean coal fires.

An entirely contrary view was formulated by James Hutton (1726–97), a Scottish chemist and small landlord, who had the money and time, and stimulus, to follow his own researches. He noted that some rocks, for example sandstones, appeared to have been made of the pieces of other rocks (such rocks are now termed 'clastic'), and he was pretty sure that slates were some sort of transformed clay – perhaps fired like a potter's earth had been fired – or, as we would now say, metamorphic. He was fascinated by the formation of gravel as rocks abraded each other in river channels. And he noted instances of soil erosion, even on small and local scales.

This led him to several revolutionary conclusions. The first is that if rocks are made of earlier rocks, then there must have been rocks before, which were worn down and deposited in the ocean. If that was the case, then current observable process had to account for what had happened, and there was no reason to suppose that prior to those rocks there not had been others, which had also been worn down, transported, and redeposited, and so *ad infinitum*. But if that happened, why was the earth not worn down once, and left flat? Obviously, new mountains had to be uplifted for the cycle to be repeated. The answer was, whatever had the strength to lift material from the bottom of the earth to the top – and clearly volcanoes could and did, and threw material miles into the air in addition. And further, he noted that volcanoes are hot; they could provide the heat to make the clay into slate. Hutton even realised that the intrusive basalt sills were not sedimentary even in their horizontal strata; they were igneous and intruded, and further proof of the forces of volcanoes.

Hutton's achievement was an epic of deduction and imagination, and of mind-blowing proportions. He had made an explanatory link between current processes, which are observable and short term, and Archaean rocks which seem everlasting and which symbolise the perpetuity of all ages. For this to have occurred, quite clearly the earth was infinitely older than anyone had thought, and Genesis was wrong. The theologians were facing a challenge as great as Darwin gave them later. Hutton said:

> But if the succession of worlds is established in the system of nature, it is in vain to look for anything higher in the origin of the earth. The result, therefore, of our present enquiry, is that we find no vestige of a beginning – no prospect of an end.
>
> (quoted in Chorley *et al.* 1964, p. 55)

God, in the form currently defined by theology, does not have to do anything. This view is very similar to those found in Hindu cosmology, but it is reached by scientific deduction and induction.

The people who supported Hutton became known as uniformitarians, or vulcanologists. They were attacked and pilloried by the establishment. The opponents' view of the earth as 'thing' rather than process was simple. The earth had been created, and was therefore not a continuous process; the soil was mantle on the earth, just like paint on a canvas. It did not come from bedrock, nor did it get removed to the sea. And as for water making valleys, what of dry wadis near the Nile, or even chalk valleys in southeast England? In short, there was no need to postulate any particular age of the earth in order to account for anything. The age of the earth happened to be however many years had passed since its creation.

As it turns out according to our current understanding, the biblical accounts were wrong, but neither Hutton nor Werner was right. Current processes do not account for all, or even much, of what we see in a particular landscape. Climatic change, and glacial advances and thaws, have produced moments of 'geomorphological terror', to be followed by longer quieter periods. Even in these quieter periods it is often the infrequent extreme event that causes most change. And the earth did have a beginning and is of an age, which we now guess at around 4.6 billion years. But in the view of the modern proponents of plate tectonic theory, the cycles of uplift and erosion can and do continue, as the earth's plates slide and grind against each other, some to be reabsorbed in the liquid interior, some to spread out as the new edges of the basal plates.

Time and the biosphere

The impact of new theories from one part of science can often be felt in cognate and far-flung disciplines, sometimes even amplified. The changing understanding of the great age of the earth was undoubtedly a key part in allowing Darwin to feel secure in his theory of the evolution of species through random mutation and natural selection. Speciation at the level of mammals – that is, the derivation of one or more new mammal species from a previously interbreeding gene pool – is not something that is 'observed' in nature, because the time-scales of mutation and selection do not conform with human life spans. What we observe in the dynamic working of nature is more often the recolonization of a forest clearing, or a volcanic hillside, with the lichens, mosses, grasses, and then more sophisticated flora and fauna of successive ecosystems, conceived as building towards a stable climax. This is all done by drawing on existing gene pools of existing organisms, even if a new web and balance may emerge between them. Thus our understanding of evolution is nearly all by inference, from observing and classifying the similarities and differences of extant organisms on earth, and by examination of the fossil record – even, with increasing skill, the extraction of some DNA sequences from the fossil record. Since modern science has given us more sophisticated understanding of DNA, and the actual biochemistry of genes, we are continually improving our understanding of rates of genetic change. We are also seeing new contests between contrasting theories of uniform or erratic speeds of speciation.

Life has existed on earth for perhaps 3.5 billion of its 4.6 billion years. For the first 2.5 billion years this life was little more than bacteria and algae. Then, at the beginning of the Cambrian geological period (600 million years ago),

> Complex life did arise with startling speed.... Readers must remember that geologists have a peculiar view of rapidity. By vernacular standards, it is a slow fuse indeed that burns for 10 million years. Still, 10 million years is but 1/450 of the earth's history, a mere instant to a geologist.
>
> (Gould 1980, p. 127)

Gould goes on to say that geologists spent a largely fruitless century trying to understand this explosion in the rate of creation of diversity. But in many ways it was pointless. There was no need to assume that diversity should be created at a constant rate in absolute time. As with compound interest leading to the exponential growth of a capital sum, so, once the variety of creatures and their complexity increases, there is a positive feedback that provides more opportunity for increasing variation.

In some senses, to the creationist school of believers in Genesis time is reversible – that is, in the sense that as the earth does not change from its preordained state as time passes, it would not matter if the sequence of days was changed and rearranged. For Hutton, time is repeatable – the endless cycle of worlds without end. But Darwinian evolution begins to raise more specific questions about the direction of time. It is taken for granted that life begins once in simple form. It is further taken for granted that with time life will exhibit more complex forms. To many human beings it also appears that there is a progression, literally meaning the making of progress, advancement, towards higher and better forms. Much of this evolution is now seen not in the singular history of each species, but in the co-evolution of whole ecosystems. Mice evolve to evade the owl; the owl evolves to catch the mice ... This suggests that there should be an inherent fragility in an ecosystem, since there will be a web of interconnecting animal and plant forms, each adjusted to the other. In practice ecosystems have been subjected to all sorts of external change, and some are more robust than others – and lower levels of life are more robust than higher.

If you want to know the direction of time, this thesis of progress in evolution, culminating of course in *Homo sapiens*, suggests that the past is simpler, the future more complex. The fossil record does not exactly support this idea. There have been several major mass species extinctions; perhaps the 'worst' was at the end of the Permian geological period, 225 million years ago. Gould (1980) explains this in terms of the changing environmental conditions as the drifting continents converge to form one landmass, Pangea. A more famous one 70 million years ago in the late Cretaceous wiped out the dinosaurs, allowing the mammals to develop rapidly in the subsequent period. The cause of this latter extinction may well have been a strike by a massive asteroid 10–20 km in diameter in the vicinity of Yucatán in Mexico, causing massive fires, polluting the air with NO_x, throwing debris into the atmosphere; all in all increasing the earth's albedo, and causing a collapse of the food chain. From the mammal line human beings develop 4 million years ago. To revert to the idea that the earth can be represented as 1 year old, man arrives at 4 p.m. on 31 December.

We, possibly the most recently separated of the mammal species, do not see speciation occurring naturally in the other higher mammals. If speciation in the higher mammals occurred fast enough for us to see it, it would be a substantial refutation of Darwinian accounts of the origin of man. If we, over the course of a few apparently unchanging human generations, could accumulate information about new mammalian species emerging, somehow we would have been selected to have a genetic stability that was not characteristic of the others. Somehow we would be special, above and beyond the rules of the game.

But as humans, perhaps we do have that ability. We have a mental capacity that is unique, and through it now the capacity to interfere with the rates of mutation and selection of genetic material. In some senses we have stepped out of time, in that no ecosystem appears to have co-evolved symbiotically with us. In other cases we have destroyed the habitats of ecosystems, threatening whole ranges of co-evolved species. On the other hand, we have also adapted ecosystems to our needs – mostly through agriculture and selective breeding – and we are acquiring the skills to design how some of the higher components should change and adapt. If humankind survives on earth, the fossil record may show apparent mass extinction and accelerated speciation occurring simultaneously, and at a rate without precedence even in the 'worst' of past geological extinctions.

One final comment is in order in this section. Though it may appear to us significant whether it is dinosaurs or mammals that rule, in Lovelock's (1979) conception of Gaia, the earth as a biological self-regulating system in which life maintains the atmosphere in the gaseous balance best suited to its own needs, the form of life is not relevant, least of all whether and in what way the higher animals exist. In many ways Lovelock's hypothesis is the ultimate homoeostatic equilibrium, and most of the changes we observe in the fossil record are just minor flips in the continuous eddy which is life on earth as a dissipative structure (see next section).

Time and equilibria

The very practical need of navigators to know their time difference from Greenwich, so that they could estimate their longitude, led to the development of very precise chronometers in the late eighteenth century. The success of these machines may have contributed to the growing dominance of an absolutist view of time. Time is precise and regular and passes of its own accord. The ticking of the clock becomes merely the symbol of what is already flowing: the clock is but a metering device for this constant flow. Such a concept of absolute time had of course proved invaluable for Newton's formulation of his laws of motion, and in turn their obvious predictive success also 'validated' and popularised the concept of time that he had used, even though in the Newtonian scheme, time is actually reversible. But we have already seen in the above forays into the history of several earth sciences that there are many complex ways in which we may think about time: for example, time is absolute/relative; or time is directional/cyclical/reversible. In this section, we will explore further some of the ways in which time can be conceptualised.

The evolutionists may point out that time is directional. The more usual statement about time's arrow is to be found in thermodynamics. For most theorists an adequate answer is given by the Second Law of Thermodynamics, which says that although energy is never destroyed, it continually changes to a state where it can do less work. A hot kettle left in a room dissipates its heat to the surroundings, and once dissipated the heat cannot be spontaneously recaptured by the kettle, to rewarm it. This is the universal trend to increased entropy – or, in statistical senses, the universal trend to disorder. Another way of expressing this, which is of great importance to a more widespread understanding of commonly unexamined assumptions, is that the trend is always towards equilibrium – where equilibrium is a kind of heat death, where there remains no further potential for change. In a relative sense, no further time passes. This equilibrium end view is one to which most classical economics subscribes. Markets will tend to equilibrium, and when prices are adjusted and supply and demand balanced, then no further change occurs. The economic signals stand stationary.

The Second Law of Thermodynamics suggests that everything is running down. In the end there will be nowhere in the universe where further work is possible. In the shorter term, however, the very flux of energy to a lower state can provide precisely the necessary conditions for the creation of higher-order forms. The theory behind this is known as the theory of dissipative structures. As low-entropy sunlight falls on earth, so winds will develop in the atmosphere, water will circulate, rain will fall, and rivers will form beds and sort their deposits. This produces an order in the landscape which we can detect. Life itself also has to capture low-entropy energy to sustain its 'improbable' form, a dynamic equilibrium far removed from the equilibrium of heat death. Life on earth captures and uses this low-entropy (negentropic) energy as it is successively reduced to high-entropy (useless) states – ultimately to background heat radiation. Schrödinger famously answered the problem of an adequate definition of life by saying simply: 'Life consumes negentropy.' Indeed, for some, the whole web of life is seen as something that maintains itself far from equilibrium by capturing negentropy. This gives a feeling for why many see life as fragile and threatened: because it is balanced so far from thermodynamic equilibrium. It also seems to imply the possibility of a 'fall' or 'collapse' – at any rate something that would happened with great rapidity. But the theory of dissipative structures suggests a contrary view: that low-entropy energy is bound to create dissipative structures as the flow to high entropy occurs.

Although the laws of thermodynamics can provide a direction for time, they say nothing about the rate of passing of time. The end is indicated, but not the duration required to reach it. Well-insulated flasks cool more slowly than poorly insulated flasks; soft rock erodes more quickly than hard rock. Similarly, in classical economics the rate at which a market will go to equilibrium is not defined. For physics as for economics it therefore becomes an empirical matter to map the changes that occur into some coordinate axis of absolute time, such as hours or years. (Actually, there is no absolute: the standard definition of time is relative to the behaviour of caesium-137.) When we do so, we find many processes have an exponential rate of decay; that is to say, they diminish more and more slowly with regard to such absolute time. An obvious example is radioactivity, where half the radioactive decay takes place in time x, a quarter (of the original amount) in the next interval x, an eighth in the next, and so on. Other (growth) processes have an exponential rate of increase.

This leads us immediately to the possibility of a reverse kind of mapping. Time is defined by events or activity occurring. If nothing happens, time does not pass. If something does happen, a unit of time passes for every equal event that occurs. The idea of equal event can of course be difficult to define – but sometimes it is not. Solar time is an obvious example: each repeated noonday sun marks another day (although the absolute length of this day increases as the earth's rotation slows up). In economics, the rise or fall of a stock by one percentage point could form the basis for a measurement of time (see below).

Chaos: or, non-equilibrial dynamics

The idea of equilibria, and specifically dynamic equilibria, has long dominated the science of ecology. Ecosystems invading a bare vocanic island will go through stages until a stable climax is reached. The rain forest regenerates itself after natural tree falls have created small clearings: equilibrium is re-established. Even when fluctuations in numbers occur, these are theorised as cyclical fluctuations. The classical example is the Volterra–Lock description of prey–predator relations – such as when sharks eat shrimps. As the sharks increase in number, the shrimps decrease. As they decrease, the sharks after a lagged period of hunger also decrease, so the shrimps increase again. This behaviour is known as periodic – in that there is repetition at a constant period. Behaviour which is fluctuating yet periodic is accepted as being in a kind of dynamic equilibrium, an end state in which no further large-scale change is anticipated.

The idea of repeated fluctuations is also present in the idea of the decomposition of an apparent random time series, by a device known as a Fourier transform. Simply put, a squiggle which seems to show no overall pattern over time, such as monthly rainfall totals in the UK for the past 100 years, can be decomposed into a sum of regular shorter and longer cycles, plus a residual 'error' term. In the case of rainfall, perhaps the shorter cycles that emerge will be annual ones, and the longer-term ones perhaps of ten years, which might represent sunspot activity.

Underlying all these ideas is the basic assumption of some kind of order, some kind of predictability, even if at the edges there are error terms too. These assumptions have, however, received a rude shock with the recent discovery and understanding of chaos theory. The theory is poorly named, since it implies the vernacular understanding of chaos as randomness. Chaos theory is, however, the theory of deterministic dynamic systems. It has been discovered by many different authors, although the paper usually cited as the first was by Konrad Lorenz in 1963, in which he published the results of his computer simulations of a simple climate. Until the advent of computers, calculations of dynamic systems were restricted to simple models, like the Volterra–Lock one. But with the advent of computers one could calculate the behaviour of systems with more than three or four non-linearly interacting component parts. It has to be emphasised that such systems are deterministic. Each state leads on to one and only one successive state. However, these systems do not go to equilibrium; they are aperiodic. Any deviation in initial conditions, even the minutest, can lead to widely divergent future paths. This is what is popularly known as the butterfly effect, that a butterfly beating its wings in West Africa may or may not set off a hurricane in the Caribbean a month later. What chaos theory says in essence is that the future cannot be predicted in any short-cut way, as can be assumed with a system going to equilibrium. The only way to know the future is to calculate it exactly from the initial conditions and the equations of change. But the catch is that the initial conditions have to be known with infinite precision, and they never can be. Complex dynamic systems are in this sense fundamentally unpredictable.

In another sense, however, perhaps there are elements of predictability. If such a system is examined in its phase space (Figure 1.1(a) and (b)), then it will be seen that there are regions of the space to which the system keeps returning, although it never returns to exactly the same point. Thus the idea that an ice age can repeat itself is not alien to such a system, nor is the idea that successive ice ages may be roughly similar. The regions that are repeatedly visited by system trajectories are known as attractors (or 'strange attractors' because the reason why they should be attractors is not well understood); these are emergent effects in different systems. If a system has two lobes to its attractor, clearly we can anticipate that it will fluctuate between two states, each of which roughly repeats previous states. But we cannot be sure, without infinite knowledge, when the system will flip from

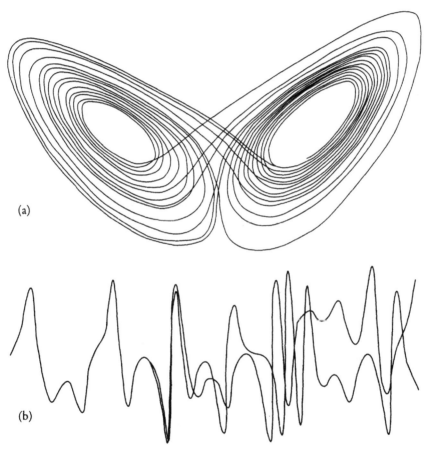

(a)

(b)

Figure 1.1 (a) The Lorenz attractor: the system has two regions within which it cycles, and randomly switches from one region to another at unpredictable moments.
(b) The butterfly effect: the two lines represent the behaviour of one variable over time (e.g. temperature) in a non-linear feedback system of several component variables (e.g. weather). The apparently insignificant difference in the two starting states results in opposite extremes of behaviour in later time periods. (From Stewart, I. (1989) *Does God Play Dice?: The Mathematics of Chaos.* Oxford: Blackwell)

one lobe to the other. So we could build a good model of the climate showing ice advances and retreats, but we would not be able to know when the next flip would occur. Chaos theory says that complex systems each have a unique history, show similar but not identical repeated patterns, are deterministic, are not predictable, and therefore show pseudo-randomness.

There is also another reason why there may be some elements of true predictability. The same system may be under the influence of a controlling parameter, for some values of which it behaves with periodicity, but for other values of which it behaves chaotically. This is exactly analogous with the heating of water: up to a certain point, nice orderly convection cells develop, but beyond boiling point the pattern of behaviour is chaotic. Just to confuse the issue further, although in this case there is a 'good progression' of behaviour with temperature, in many systems the distance between the values of the control variable which cause periodic or chaotic behaviour is arbitrarily small, and the sequence of values of the control variable which cause one or the other is not a 'good progression'. This would be the equivalent of saying that as temperature increased by the minutest amount at any point between 0° and 100 °C water would seem to go arbitrarily from convective to boiling patterns, and vice versa.

TIME AND THE SOCIAL WORLD: CAUSATION, PAST AND FUTURE

In the natural sciences teleological explanation is by and large prohibited. That is, explanation based on the attainment of a future goal is not deemed scientifically acceptable. The eye of the animal does not evolve in order that the animal may have vision, neither does the ear evolve in order that the animal might have hearing. Effects are only allowed to follow cause, and never the reverse.

From the point of view of the social sciences, this poses a problem, summarised by Tolstoy in the following manner:

Reason says: (1) Space with all the forms of matter that give it visibility is infinite, and cannot be imagined otherwise. (2) Time is infinite progression without a moment's pause, and cannot be imagined otherwise. (3) The connexion between cause and effect has no beginning and can have no end.

Consciousness says: (1) I alone am, and all that exists is only I; consequently I include space. (2) I measure flowing time by the fixed moment of the present, in which moment I alone am conscious of myself as living; consequently I stand outside of time. (3) I am independent of cause, since I feel myself to be the cause of every manifestation of my life.

Reason gives expression to the laws of necessity. Consciousness

18

gives expression to the reality of free will.

(Tolstoy 1869/1957, p. 1438)

Reason, here synonymous with science, sees the world as one web of deterministic cause and effect. This even applies to the idea of evolution, where each random variation of genetic material in principle can have a deterministic cause. But the supreme characteristic of being human is to be self-conscious. That self-consciousness gives rise to the idea of a future, and of free will, and hence the possibility of teleological explanation. Mostly the non-fatalists among us believe that we have choices, and that we can select that choice which will best achieve our future ends. The human distinction between the past as inevitable and the future as boundless possibility is the intuitive way of understanding time's arrow, whether or not ultimately the very concept of free will proves illusory (as Tolstoy believed).

All theories of social change have to struggle somehow around this pivot. Some may stress the supremacy of social law and determinacy, others may celebrate human agency and free will, some may try to strike a continuous reciprocal balance between the two. In this section we will examine briefly some theories of social evolution. But it is important to realise from the outset how in nearly all of them the natural world – let us call it the environment – plays virtually no role. It is left out either because it is not seen as changing – it is a backcloth – or else because it adds a dimension which complicates the time-scales with which theory deals.

Economic development

Theory in science is continually tested empirically. Theory in social science may start from empirical foundations, but often becomes normative and prescriptive, and to that extent can often simplify or choose its own time-scales. Karl Marx's materialist theories of history demoted the significance of the individual, and saw in the economic relations of production of the capitalist state a mechanism that would so concentrate wealth, and so widen and deepen poverty, that it would inevitably lead to revolution and the final establishment of the communist society. Change in this system is therefore inevitable – but no time framework is put on it. As with thermodynamics, we have direction, but not the rate. It is a stage model, one in which the stages are ordered and which therefore act as time's arrow, but yet the speed with which the stages will pass is unknown. The best that Marx could see was that the process could be hastened by the cultivation of the class consciousness of the proletariat. In the 1960s at the height of the Cold War, when the USSR's transformation from backward agrarian to modern industrial state in little more than forty years was seen by many as remarkable, and during the period of decolonisation of Africa and Asia when both Western and Eastern blocs were vying for Third World allies, the American economist W. W.

Rostow published his *Stages of Economic Growth: A Non-communist Manifesto* (Rostow 1960). The subtitle is often forgotten, but it was quite explicit, because most Western economists had simply assumed that capitalism and economic growth went hand in hand, and few had thought of how a new government might want instruments to foster development. Rostow's model established the word 'take-off' in the international diplomatic lexicon, a fundamental prerequisite to the establishment of the International Development Decade and the promotion of international aid for the Third World. With the establishment of the word came the belief that a once-and-for-all transfer of resources from the First (and Second) Worlds to the Third would kick-start the less developed economies. If ever there were a word laden with temporal implications, it is 'take-off'. This is a critical moment, when a build-up of speed lifts something from one world to another, where it will suddenly soar higher and higher. The temporal framework of Rostow's model was again, like Marx's, essentially a simple one of stages. To be sure, there are instances of an understanding of time – in the stage following a timeless traditional society agricultural reform would enable domestic savings rates to rise – but Rostow did not imply that he knew how fast the Third World would or could change. Thus in neither Marx's nor Rostow's model is it possible to say how long development will take in terms of any other time framework. Nevertheless, the politicians on both sides of the North–South gulf picked up Rostow's provocative term and thought that a development decade would at a stroke speed everyone to 'take-off.'

Other development economists incorporated time more specifically in many of their models. In the Fei–Paauw and Chenery–Strout models, the key variables include rates of investment, and rates of savings and capital/output ratios contrasted with other rates of change – specifically and importantly, rates of population growth. Capital accumulation cannot occur nor wealth increase unless the former are favourable in comparison with the latter. Interestingly, the rate of change of these variables with respect to clock time is not important. A very low savings rate can in time result in the same rate of wealth accumulation as a very high savings rate, if the former is accompanied by low population and the latter by high population growth rates.

However, any economic model implies inputs and outputs from the environment. These too will occur at differing rates, and in turn be related to differing rates of environmental change. These rates may include absorption of waste, or biomass renewal, or other cycles such as the atmospheric cycles of water vapour or trace gases. In general our modelling of these is weak, and our modelling of the interrelationships between the human and physical systems even weaker, but this has not stopped attempts to simulate these interactions, an early and famous case being *The Limits to Growth* by Meadows *et al.* (1974). The problem with such models is that they tend to one of two results: either they crash to an end state (a boundary) in which one variable dominates (pollution kills everything), or else they behave chaotically, in which case we know that small shifts in initial conditions result in arbitrarily large divergencies in predictions. Nor are they able to incorporate changes in technology which have yet to be invented.

The complexity of change

What time is relative to depends of course on the specific centre of attention of the human observer. To an analyst of the short-term behaviour of stock and shares, time passes in the stock-exchange only when the stock-exchange is open for business. The hours of the day and the days of the week when it is shut are discounted as non-existent time, even though for the analyst in her role as parent this other time does have a meaning. Whether for one person his involvement in many different time frames is a source of stress or a source of relief is for the moment conjectural. There is no doubt that for some people rapid change in one part of their life can be accepted only if there is a counterbalancing stability in another part of their life. This is something which the ministers of religion understand well, with the notion of unchanging ritual in patterns of worship.

The idea that time is relative to events occurring explains something of our attitude towards the 'speed' of contemporary human life. A few centuries before the industrial revolution most people of most generations did not expect to see much technical change in their own lifetime, and expected that their children would follow similar occupations using similar methods to themselves. To such a person the periods of time of interest in their own lives were the daily and annual rhythms, and two entropic times: first, the stages of their own body, Shakespeare's seven ages of man, from youth through puberty and marriage to senility and death, and second, the succession of the generations. But in our contemporary life the number of successive technical changes is such that we have an experience of time passing very fast, for example from external to internal combustion, and to nuclear fission; from handwritten letter and horse, to telegraph and wire, to telephone, fax machine, modem, and e-mail. Most of us can remember past decades, the 1960s or 1970s, as a long time ago because of the changes wrought since then. We also have an increasing number of reporting technologies and agencies specifically dedicated to the idea of informing us of change – or what one might call 'news'. In mass media terms, news generates a high-frequency expectation of low-frequency events. Wars and aircraft crashes are not commonplace, nor are floods and earthquakes. But our expectation of seeing or hearing daily of such an infrequent event is actually very high.

The human experience of the environment in a way might follow similar logic. The timeless and ageless rain forest is timeless and ageless if, and only if, it does not change, and is then not news but natural history, and implicitly in some time-independent equilibrium. If on the other hand at one site we

21

observe a great deal of environmental change, do we develop a sense of urgency, that somehow environmental change is happening too fast, faster than it ought to happen? Do floods in Bangladesh mean the erosive collapse of the Himalayas? There is evidence that we do just that, despite the fact that in many situations landscapes are not near equilibrium, that man has negligible impact on many large-scale processes, and that some local geomorphic processes happen very quickly in human terms – witness cliff collapses and retreats, mudslides and changes in distributary patterns in deltas.

Finally, we enter the political domain, which is predicated on the idea that we do have free will and we do face choices – whether these are taken undemocratically or democratically. In this domain the environmental knowledge of experts trained in the natural sciences collides with the expectations and perception of the broad public, mediated by politicians working to a timetable set by constitutional and career requirements. Governments have only so much of a lifetime, and the politicians within them only their own professional lifetime in which to climb the political tree. The experts may warn us of long-term dangers. The public may be unable to comprehend them – and possibly be rightly sceptical. The politicians in any event may be able to think of the distant future, but be able to act only in the short term.

Conversely, it is in the interest of some political actors, specifically environmental pressure groups and other environment-related non-governmental and intergovernmental organisations, to stress the immediacy of potential environmental destruction. If they are to maintain, or increase, their profile (and attract funding) it is in their interest to stress that they are attacking important and immediate environmental issues. This is not to say that pressure groups deliberately exaggerate the rate of environmental change; but the neccessity of maintaining public interest inevitably leads to the choice of particular issues or images. The rapid clear-felling of tropical rain forest in the Amazon provides a more striking image than the gradual process of desertification on semi-arid lands, a fact reflected in the respective amounts of coverage each receives in international environmental campaigns (unless, of course, there happens to be a famine in a particular semi-arid region).

THE CONTENTS OF THE BOOK

This is not the place to comment on the contents of each chapter; we make some introductory remarks as a preface to each in turn. Here we will restrict our remarks to some comments on why the material appears in the order that it does. The order has to be arbitrary since this is multi-dimensional material – several chapters cutting across many themes – so the one-dimensional arrangement of a book is a necessary but awkward constraint.

Our initial thoughts were to start with longer-term environmental change, say at the millennial scale, work through the century scale, and come to the finer-grained chronologies towards the end. To some extent we have done that, and to some extent this means that the chapters with more physical geography in them occur at the beginning. Therefore the chapters more concerned with environmental science are found towards the beginning. But this is actually to simplify too much. Neil Roberts looks critically at current human use of the depressions of the plateaus of Southern Africa known as dambos, which may be relict features of the Pleistocene. Jean Grove addresses the question of how fast medieval society adjusted to changing climate in Greenland and England – for example. Her sources of evidence are multifarious, from field observation and historical documents as well as instrumental records in the more recent past. Nicholas Clifford and John McClatchey look specifically at the instrumental record of temperature change in England, and show how hard it is to be able to say anything certain about climatic trends for even one variable.

In Chapter 5 we move from past to future, and to the marriage of the approach of the atmospheric scientist to that of the economist – seeking to find the most reasonable way of evaluating future impacts against current controls on emissions of greenhouse gases. The fairly simple comparison of the time-scales of the scientist and the economist then comes up against Taplin's account of the reality of the differing time-scales for action acceptable to politicians internationally, to politicians nationally, to the public, and to the media. There is a lucid sense conveyed of 'muddling through' and ultimately of 'fudge'. John Gordon provides a comparison with the case in the UK, where he too finds that the time-scales for impacts, policymaking, implementation, and public opinion formation simply do not marry well enough.

The complicated way in which society sees its environment through the lenses of its own political economy is well developed by William Beinart, who looks at the history of rangeland stocking in South Africa. He shows quite clearly that change does not have to mean degradation, that change is not linear, and that the perception of the history of land management itself changes with the history of the society. This is important because, as Melissa Leach and James Fairhead show, to presume that someone standing in a poorly forested land showing signs of denudation is culpable of mismanagement, is to presume that the observer knows which way the landscape is changing. This may be the remnant of a golden past heading for a catastrophe, or it may be the current first stages of recapturing and ameliorating hostile badlands.

Many of these chapters have thus touched on the complex ways in which different things – temperature, rainfall, legal systems, capital accumulation, political systems – change with differing speeds. There is a question here: can we make theoretical sense out of these different time-scales, and can we make useful predictions on which to base current action? Malte Faber and John Proops address these problems from a conceptual angle: they do not necessarily find answers, but they illustrate the complexity of the approaches necessary to begin to find answers, and also the limits in some areas to our abilities to predict. We close with a chapter on public opinion formation, media, development, and environment in India. The chapter asks questions about what the public in India know about environmental issues, and what they are prepared to do about them. By and large their perceptions are local and short term. To alter public opinion will take a much larger effort than currently is taking place, and over a generation or two. In the case of India, even then environmentalism might not grow wider roots until the level of development is raised, poverty eliminated, and population growth vastly reduced – and on current evidence all of that operates on time-scales which completely frustrate the urgent 'now' demanded by the Northern environmental movement.

REFERENCES

Chorley, Richard J., Beckinsale, Robert P., and Dunn, Antony J. 1964. *The History of the Study of Landforms, or, The Development of Geomorphology.* Vol. 1: *Geomorphology before Davis.* London: Methuen.

Cohen, Jack, and Stewart, Ian 1994. *The Collapse of Chaos.* London: Penguin.

Gould, Stephen Jay. 1980. *Ever since Darwin: Reflections in Natural History.* London: Penguin.

Lorenz, C. 1963. Deterministic nonperiodic flow. *Journal of the Atmospheric Sciences* 20: 130–41.

Lovelock, J. E. 1979. *Gaia: A New Look at Life on Earth.* Oxford: Oxford University Press.

Meadows, Donella H., Meadows, Dennis L., Randers, Jørgen, and Behrens, William W. III. 1974. *The Limits to Growth: A Report for the Club of Rome's Project on the Predicament of Mankind.* London: Pan Books.

Mills, C. Wright. 1963. *The Marxists.* Harmondsworth: Penguin.

Ornstein, Robert E. 1969. *On the Experience of Time.* London: Penguin.

Rostow, W. W. 1960. *The Stages of Economic Growth: A Non-communist Manifesto.* Cambridge: Cambridge University Press.

Tolstoy, L. N. 1869/1957. *War and Peace*, trans. Rosemary Edmonds. London: Penguin.

Whiteman, Michael. 1967. *Philosophy of Space and Time.* London: Allen and Unwin.

Yeats, W. B. 1937. *A Vision.* London: Macmillan.

2

LONG-TERM ENVIRONMENTAL STABILITY AND INSTABILITY IN THE TROPICS AND SUBTROPICS

Neil Roberts

Editors' note This chapter provides a useful synopsis of our understanding and recognition of the history of the Earth's climatic and landscape change over the last million years, though concentrating on the last few hundred thousand years. Some of Neil Roberts's findings should be recalled as later chapters are read. Changes in solar radiation due to the earth's orbital variations correspond quite well with successive ice advances, but they may have been enhanced by the changing atmospheric concentration of greenhouse gases, which also correlates well with temperature change in the last 150,000 years. This can be borne in mind when reading Max Wallis's chapter (Chapter 5) on costing global warming. Roberts's conclusion that changes in precipitation at low latitudes are as important as changes in temperature at high latitude and can occur quite rapidly, and that landscapes are quite often out of equilibrium with current conditions, is of significance to a better understanding of William Beinart's chapter (Chapter 8) on the perception of environmental change in Southern Africa.

INTRODUCTION

Environmental change, whether the result of natural (e.g. climatic) or human factors, is an inherent feature of all earth surface systems, but different processes of environmental change operate over a range of time-scales. For example, some organisms (e.g. insects) can respond to changed environmental conditions within a matter of months, whereas most tree species require several decades to become established and grow to maturity on previously vacant land. Other processes, like soil formation, are likely to take place over centuries or even longer time periods. Because of this, a range of different methods is needed for the study of environmental change. Over time periods of a few decades or less, the methods of observation and monitoring can be employed, either through direct measurement or via remote sensing. Satellite imagery is proving to be an especially vital tool in monitoring global and regional changes in the extent of ice and snow, in vegetation cover, and in oceanographic conditions.

However, many environmental systems change over time-scales longer than those which can be observed directly, and it is therefore necessary to turn to human and natural archives. Over intermediate time-scales (viz. decades to centuries), documentary sources of data can be used, although these data were not always collected in a form ideal for the analysis to which we wish to subject them. The length of the 'historical' time period also varies greatly in different parts of the world. Whereas we have written records from ancient Sumer, including data on early problems of land degradation through salinisation, back to 3000 BC, prehistory in the Papua New Guinea highlands ended only in the 1930s. These sources have provided extremely important data concerning changes in climate and human impact during the last millennium. Grove's work (see next chapter) on the Little Ice Age climates in Scandinavia, the Alps, and Crete, as derived from tax records and other documents, provides an excellent example of this approach.

Many low-latitude regions, such as inter-tropical Africa, have relatively short archival records, however. Here, and for Holocene or longer time-scales, proxy methods of palaeoenvironmental reconstruction become essential. The specific methods involved are numerous and can be found described in detail elsewhere (Lowe and Walker 1984; Bradley 1984; Roberts 1989). Among them are:

- tree-ring analysis (tree-ring widths are linked to weather conditions; this technique has high resolution and precise dating, but there are few data from outside Europe and North America);
- pollen analysis (as an indicator of past vegetation and hence climate and/ or human impact; this is now widely applied as a technique);
- lake-level and salinity fluctuations in closed-basin lakes, associated with changes in water balance over the lake catchment (reconstructed from lake sediments and the fossil organisms, e.g. diatoms, which they contain);
- geomorphological records, for example of fossil sand dunes found beyond the limits of modern deserts. Like lake-level data, these are especially important in dryland areas of the world;
- stable isotopes of elements such as oxygen and carbon, which reflect – among other things – past global ice volumes and changing temperatures. These have been applied particularly to deep-sea sediment cores, but are also increasingly used in continental sequences.

Analyses such as these need to be placed in a time frame and hence require dating. A wide range of dating methods now exist but most fall into one of two categories:

- based on natural increments; e.g. tree rings, laminated (varved) lake sediments
- radiometric; usually based on the decay or unstable isotopes, such as carbon-14 or uranium–thorium dating.[1] Another, related dating technique is based on different types of luminescence.

In addition to these approaches to reconstructing environmental change is the use of numerical modelling. Computer-driven simulations of the global climate system involving general circulation models (or GCMs) provide the basis for predictions about the future response of climate to enhanced greenhouse-gas forcing. All the main GCMs are able to replicate the earth's present climate with considerable success (Mitchell and Zeng Qingcun 1991), but they produce significantly different projections for the future, notably under the scenario of doubled CO_2 equivalent. While all GCMs predict some warming, estimates differ between +1.9 and +5.2 °C. For other factors, such as effective soil moisture, there is even less agreement, with individual regions predicted to become wetter in some GCM experiments but drier in others. According to the United Kingdom Meteorological Office (UKMO) model, a doubling of atmospheric CO_2 will cause precipitation in the tropics to increase by an average of about 15 per cent but, despite this, there will be an absolute decrease in soil moisture status because of higher evapotranspiration.

In fact, it is essential to use these different approaches in combination. For example, in the case of numerical models what we need to know is how reliably they simulate climate *change*; that is, indicate with confidence conditions significantly different from those of today. For this, different GCMs need to be tested against data sets other than the present-day climate in order to calibrate the models. One approach which has been adopted has been to test them against data sets based on past climates; a comparison between numerical simulations and reconstructed climatic data since the last glacial maximum formed the basis of the successful COHMAP project (COHMAP members 1988). Comparison of different types of record also allows a cross-check to be made on the reliability and sensitivity of different methods. For instance, there is a good historical record of water-level changes in Lake Naivasha, Kenya, over the past century linked to changes in rainfall over the catchment. These have been compared to lake salinity in an embayment of the lake as reconstructed from invertebrate remains in a short sediment core (Verschuren 1994). The two show a good match, giving support to the 'proxy' method and allowing it to be extended to longer time periods with greater confidence.

The focus of this chapter is on long time-scales using 'proxy' methods to reconstruct past environmental change in the tropics and subtropics. It is important to consider these longer time-scales if we want to establish how typical present-day rates of environmental change are, compared to those of the past, and – where there has been major human disturbance – to help establish what the natural pre-disturbance condition of the environment was like. I shall start with climatic changes over tens of thousands of years, and then move towards the present day to make some brief reference to the role of human impact on landscape transformation.

MILANKOVICH TIME-SCALES OF CLIMATIC CHANGE

Major long-term climatic changes in the tropics and subtropics took place over time-scales similar to, although probably not identical with, those of high-latitude glacials and interglacials. Oxygen isotope analysis of deep-sea cores has provided an idea of changing global ice volumes (i.e. glaciation) and has shown that full glacial–interglacial cycles occurred about every 100,000 years during the past million years. This corresponds to the longest of three principal astronomical cycles (the orbital eccentricity) recognised by Milutin Milankovich as influencing the earth's receipt of solar radiation (Berger *et al.* 1984). On the other hand, orbital forcing in itself seems unlikely to have induced major changes in climate; it would have required interaction with other mechanisms. One of these may have been the atmospheric concentration of greenhouse gases, which ice-core data show have oscillated in close harmony with global temperature during the past 150,000 years (Lorius *et al.* 1990).

In the tropics and subtropics, oceanic and other evidence suggests that the two shorter orbital cycles (axial tilt 41,000 years, precession 22,000 years) may have been more important in modulating climatic change. The record of atmospheric methane preserved in ice-cores shows a strong oscillation corresponding to these cycles, and suggests that the main source of variations in this greenhouse gas during the Quaternary may have been changes in the extent of tropical wetlands, such as the Sudd swamps of the Nile (Petit-Maire *et al.* 1991). Although temperatures did fall in the tropics during glacial stages by up to *c.* 6 °C, an equally large change was wrought by changes in rainfall, and the main palaeoclimatic signal is consequently change between humid and arid conditions rather than between cold and warm. This has been linked primarily to the location and intensity of tropical atmospheric circulation systems, such as the monsoon, and has led to large changes in the extent of the world's drylands during the past 20,000 years.

It has long been recognised that the Saharan, Arabian, and Rajasthan deserts were not always as they are today (Figure 2.1). Rock paintings from the Tibesti mountains of the Sahara depict herds of cattle, gazelle, and giraffe, and suggest a well-watered land which, like the modern savannahs, was rich in animal life. At Oyo in the heart of the eastern Sahara, pollen was found of plants that are today found 500 km to the south (Ritchie 1994), and other pollen diagrams confirm that the northward shift of tropical vegetation belts

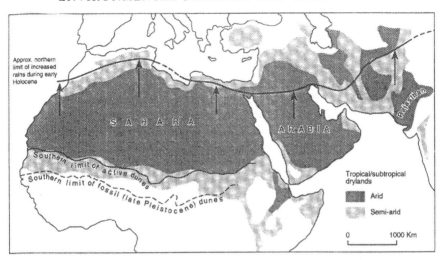

Figure 2.1 Changing climate of the Saharan–Arabian dry zone

occurred over a broad sector of the Old World tropics and subtropics (Singh *et al.* 1974; Lézine 1989). Lakes and their sediments provide one of the best lines of evidence for the changing climate and hydrology of the world's drylands. Lacustrine deposits are found in such presently arid locations as the Rub'al Khali of Arabia and the Taoudenni oasis of northern Mali, and testify to times when the water balance was more favourable than it is today. Palaeosalinity indicators, such as diatoms, show that many lakes which today are saline formerly contained fresh water, and were able to support rich and varied aquatic faunas.

On the other hand, there is evidence of conditions considerably drier than at present in many areas of the tropics. The clearest indication of this comes from sand-dunes which are no longer active, but lie outside the modern desert. Because of the close relationship between aeolian processes and climatic aridity, fossil dunes help to provide an indication of the changing extent and location of the world's arid zone. On the southern (Sahelian) margin of the Sahara, active sand-dunes are today restricted to land with less than 150 mm mean annual rainfall. Stable, vegetated dune ridges are none the less found up to 450 km south of this limit, in areas presently receiving up to 1,000 mm rainfall (Grove and Warren 1968). These mark a former extension of the Sahara southwards, when there was a reduction of effective precipitation to only a quarter of its present value. Fossil sand-dunes are also found in Australia and northwest India, and they have even been traced beneath the tropical rain forests of Zaïre and Amazonia (Goudie 1983).

It was once believed that wet–dry cycles in the tropics could be correlated with the glacial–interglacial fluctuations in temperate and polar regions – the cold periods of high latitudes being matched by 'pluvial' periods in lower latitudes. In fact, radiometric dating has shown that in the tropics, arid rather than humid conditions coincided with the period around the last glacial maximum (20,000–15,000 cal. yr BP[2]). The last wet period occurred after this, during Northern Hemisphere deglaciation and through the first half of the postglacial – or Holocene – period (14,500–6,000 cal. yr BP). At this time the Sahara and Arabia could hardly be called deserts at all, and precipitation estimates for East Africa, the Sahel, and South Asia indicate an average annual increase in rainfall of at least 250 ± 130 mm (Street-Perrott et al. 1991). It is clear that the wetter climate of the early Holocene was brought about by an intensification and northward displacement of the monsoonal circulation, particularly the South Asian monsoon. However, these monsoon rains do not appear to have reached most of the Mediterranean region with any regularity. In Turkey and western Iran pollen and lake levels suggest a drier rather than a wetter climate than at present (Roberts and Wright 1993), and the same is true of some other extratropical regions like the American Mid-West.

The early mid-Holocene period has been believed to incorporate an interval known as the climatic optimum or Hypsithermal, when temperatures may have been 1.0–2.5 °C higher than today. This has been of particular interest because of its potential similarity to a future world warmed by anthropogenic increase in greenhouse gases. Butzer (1980) and other authors, using this as an analogue, have suggested that greenhouse gas warming may lead to an increase in rainfall in low-latitude areas such as the Sahel, and the desiccation of mid-latitude regions including the grain-producing areas of the American Mid-West and the former Soviet Union. Such a change would, of course, have major implications for the economic and geopolitical balance between the South and the North. In reality, the enhanced summer rains of the early mid-Holocene were brought about by enhanced seasonality and continentality caused by orbital changes, notably in the 22,000-year precessional cycle. In contrast, the radiative forcing caused by greenhouse gases causes mean annual temperatures to rise, but with little change in seasonality (Mitchell 1990).

At the beginning of the Holocene, the Northern Hemisphere received almost 8 per cent more solar radiation during the summer months than it does today, although correspondingly less in winter. Because the strength of the monsoon circulation is a function of heating of air over landmasses, the higher summer temperatures over the Northern Hemisphere at that time led to more intense and widespread monsoon rains in the northern tropics. GCM-simulated values for effective moisture (precipitation minus evaporation) show good correspondence with the lake-level record during the past 20,000 years (Kutzbach and Street-Perrott 1985). On the other hand, a lag is observed at 7,000–6,000 cal. yr BP, when seasonality was much weaker than earlier in the Holocene, but when tropical climates remained relatively wet. Other factors were apparently able to override the influence of astronomical forcing during this transitional stage. In this case the persistence of well-vegetated land surfaces in regions such as the Sahara may have maintained albedo values at sufficiently low values to extend the early Holocene wet period beyond its allotted time span (Street-Perrott et al. 1991). This highlights the importance of linkages between climatic conditions and vegetation cover at the regional scale, and suggests that a large-scale change in ground surface albedo brought about by contemporary desertification might also be capable of contributing to a decline in precipitation, such as has been observed in the Sahel during the past two to three decades. (Incidentally, one of the reasons why the Sahel has been so prone to desertification is that it once was desert and the fossil sand-dunes can easily be reactivated by the wind once the vegetation cover is stripped off by livestock or removed for wood fuel.)

SUB-MILANKOVICH TIME-SCALES OF CLIMATE CHANGE

The tropical monsoon circulation was, in summary, weaker than at present around the time of the last glacial maximum, but more intense during the early Holocene. However, detailed analysis of palaeoclimatic records during the last glacial–interglacial transition indicates that neither temperature nor rainfall increased steadily in line with the gradual shift in the earth's radiation balance after 20,000 cal. yr BP. For example, warming was interrupted between 13,000 and 11,000 cal. yr BP by the Younger Dryas event, when temperatures fell and then rose again by >1 °C per decade. An equally abrupt change in precipitation is recorded in African lake levels at this time (Gasse et al. 1990; Roberts et al. 1993). Feedback mechanisms thus appear to have caused the climatic response to astronomical forcing to have been non-linear in character. These abrupt shifts between arid and humid phases often created a temporary disequilibrium between climate, biotic processes, and geomorphic processes, when rainfall erosivities were high, but a protective vegetation cover had not been fully established.

In inter-tropical Africa, the last important dry-to-wet climatic transition occurred c. 14,500 cal. yr BP at the end of the late-Pleistocene arid phase. Data on sediment accumulation rates in lake cores (Roberts and Barker 1993) and from alluvial stratigraphies in river valleys (Thomas and Thorp 1995) indicate that catchment erosion and sediment flux peaked immediately after this dry-to-wet shift in climate. This accords with Knox's (1972) biogeomorphic response model and highlights the role that vegetation plays in mediating the erosional response to climate. This period of climatic transition may provide a valuable means of calibrating the future response of tropical landscapes to the climate projected to occur under a greenhouse-gas forcing.

The present-day climatic regime was established in low-latitude regions during the mid-Holocene, c. 6,000–5,000 cal. yr BP, immediately prior to the construction of the pyramids in Egypt and the emergence of several Old World civilisations. Although historical times have thus far been characterised by a single, dominant climatic mode, there have been significant departures from it at a regional and possibly also global scale. At low latitudes the most important departures have been in the form of abrupt arid events, comparable in amplitude and duration to the Younger Dryas stadial (Gasse and Van Campo 1994). Several of the same tropical lakes which recorded a sharp fall in water level at the end of the last glaciation also experienced important phases of negative water balance around 8,300, 5,300, and 4,100 cal. yr BP. Furthermore, lake-level data from North Africa indicate that these 'abrupt events', which lasted at most a few centuries, affected not only the tropics but also the Mediterranean zone (Gasse *et al.* 1990; Lamb *et al.* 1995). This synchronicity between climatic trends on each side of the Sahara stands in contrast to both short-term twentieth-century trends and those over long-term Milankovich time-scales, when dry and wet phases do not appear to have been in concert between regions of winter and summer rainfall (Roberts *et al.* 1994).

Data are beginning to emerge which suggest that tropical climatic 'crises' may have occurred in recent centuries as well as in recent millennia. In particular, sediment cores from Malawi in East Africa indicate a fall of c. 100 m in the water level of this great lake within the last one thousand years, signifying a reduction in rainfall to 50–70 per cent of modern values at or around the time of the Little Ice Age (Owen *et al.* 1990). The causes of these perturbations to the climate system are not certain, but among the possible explanations are major volcanic eruptions, variations in cosmic ray flux in the upper atmosphere, and oceanic mechanisms such as the switching on or off of North Atlantic Deep Water (NADW) formation (Street-Perrott and Perrott 1990). Even apart from these major perturbations, it seems that the tropical climate has been inherently variable, if not unstable, with oscillations between periods of drought and wetness at a hierarchy of different time-scales during the late Holocene (the past 2,000–3,000 years). Data are still too sparse to provide anything like a complete record of climate changes; indeed, in the Old World tropics and subtropics they are notably poorer for the past thousand years than they are in Europe. Rectifying this deficiency is one of the goals of the current PAGES (Past Global Changes) programme of the International Geosphere–Biosphere Programme (IGBP). None the less, a range of archival and proxy data indicate that periods of extended drought, such as has been experienced by the Sahel since 1970, or unusually wet years, such as affected East Africa in the early 1960s, are by no means unique on a time-scale of millennia. Nicholson (1980) has shown that the semi-arid zone of West Africa was wetter than at present from the sixteenth to the eighteenth centuries, but was dry for much of the nineteenth century. Her reconstructions were based on numerous records but individually they were discontinuous or of low resolution. A higher-resolution sequence comes from climatically calibrated tree-ring data from Morocco, where variations in moisture would have been related to winter precipitation (Till and Guiot 1990). Thirteen periods of drought were identified for the period since AD 1100, with a maximum duration of six years, the most severe occurring c. AD 1200.

HUMAN IMPACT ON ENVIRONMENTAL CHANGE: THE CASE OF DAMBO HYDROLOGY

One of the most important aspects of environmental change is its effect upon the landscape. Human agencies have caused major transformation of 'natural' landscapes over decadal to millennial time-scales, and these effects have generally increased towards the present day. In terms of proxy data, human impact on the landscape is best recorded by pollen-based records of deforestation. Loss of vegetation cover has typically led to accelerated soil erosion, and this is reflected in increased rates of sedimentation in river valleys and lakes. The timing of this increase in sediment flux differs widely from one part of the world to another. In much of Europe and the circum-Mediterranean lands, this transformation from natural to cultural landscapes occurred several millennia ago, whereas in much of the tropics it appears to have occurred substantially later in date. For example, despite contemporary evidence of accelerated erosion, most sites in sub-Saharan Africa do not show the up-core increase in sediment yield typically found in Europe. This suggests that human-induced acceleration in erosion is of very recent origin in most of Africa, probably within the past hundred years.

The regional consequences of human impact and climate change have often been complex, and it may be hard to disentangle their respective influences. Even so, it is important to establish this baseline, if we are to establish correctly the long-term sustainability of any particular forms of environmental resource use. I shall use the effect of changing climate and forest cover in the catchments of dambo wetlands in Southern Africa as an illustration of this.

Dambos (also known as *vleis, mbuga,* or *fadamas*) are shallow, seasonally waterlogged depressions found near the head of drainage networks, and are common in Africa's tropical plateau savannas (Mackel 1985; Roberts 1988). They are generally treeless with the vegetation dominated by grasses and sedges, but the surrounding catchments are normally covered by dry (e.g. miombo) woodland under natural conditions. Individual dambo wetlands are typically small: 0.1–1.0 km wide and 0.5–5.0 km long. Because they remain moist during the dry season, dambos represent a reliable, near-surface source of water for human and animal consumption, for garden cultivation using small-scale irrigation, and for grazing of livestock. There is a long tradition of simple irrigation in Southern Africa, stretching back into the pre-colonial era, as evidenced from early travellers' reports of abandoned field systems. The extent of this informal garden irrigation has increased significantly in many parts of Africa during recent decades (Bell and Roberts 1991). On the other hand, there has been active debate about the sustainability of different forms of dambo utilisation. Some dambos show signs of environmental degradation (Whitlow 1985), and fears have been expressed that cropping of wetland lowers the water-table and reduces dry-season stream flow, that it renders the soil infertile by oxidation of organic matter, and that overgrazing increases the hazard of soil erosion by both gullying and sheetflow (Faulkner and Lambert 1991; Roberts and Lambert 1990). However, erosion and desiccation are the products of a complex chain of events linking both human and environmental factors over a range of time-scales (Bell and Roberts 1991).

Dambos have a core area in central Southern Africa where mean annual rainfall is in the range 800–1300 mm. It has been suggested (e.g. by Bond 1967; see also Whitlow 1985, p. 119) that where mean annual rainfall is below this, dambos may be 'inactive' and hence represent a non-renewable resource. This would include, for example, many of the dambos in Zimbabwe, where they cover 1.28 million ha (3.65 per cent of the land area), and where they are an important resource to peasant farmers in Communal Areas. It seems likely that these dambos would have been 'activated' hydro-geomorphologically only during times of higher rainfall, such as the early Holocene (Figure 2.2). At that time the whole of the dambo would have been wet, at least seasonally. With subsequent climatic desiccation, however, water-tables would have been lowered and the dambos would have, to varying degrees, dried out, particularly in the dry season and in their lower parts.

In more recent times, however, they appear to have been partially reactivated, inadvertently, by human activity. Evapotranspiration by trees in the catchment causes a significant lowering of water-tables and a reduction in throughflow to the dambo (Faulkner and Lambert 1991). With gradual deforestation and the replacement of woodland by dryland maize fields, groundwater yields to the dambo increased once again, and the wet area of the dambo expanded. Replacement of grasses and sedges by irrigated garden crops seems to have had little overall effect on evapotranspiration losses. On the other hand, clearance of woodland has substantially increased the proportion of water being lost by runoff, rather than by throughflow, and this is lost more quickly from the dambo system. Runoff, combined with heavy grazing pressure, has also undoubtedly caused the gully erosion found in many of the 'drier' dambos today. Establishing what is a sustainable pattern use of dambo resource is thus complicated by the fact that dambo hydrology has not been constant, but has shifted through time as a complex response both to climatic change and to human impact.

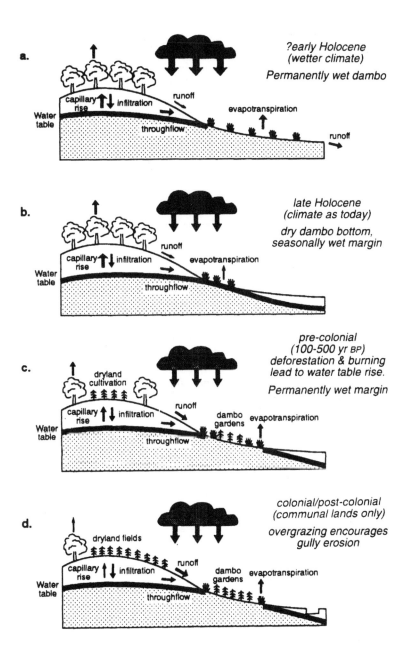

Figure 2.2 Changing dambo hydrology under different scenarios of climate and human impact

CONCLUSION

The record of past changes helps us establish the long-term sustainability of environmental resources and their usage, and also provides us with clues about the environmental responses to future climatic change.

First, palaeoenvironmental records and numerical models both indicate that the most critical changes in future climate in low latitudes will relate to water balance, especially moisture stress and drought, rather than purely to warming. Palaeoclimatic data show that large changes in the extent of the world's drylands were associated with relatively small changes in global mean temperature. However, neither historical data nor general circulation model (GCM) experiments are able to predict with any reliability the magnitude, or even the direction, of future changes in water balance at a regional scale (i.e. at a scale useful for establishing policies to mitigate, or to benefit from, the consequences of climate change). There is disagreement between different GCMs, while past climates do not offer appropriate analogues for future greenhouse-gas modified climate.

Palaeoclimatic records do, however, indicate that periods of major climatic transition in the past were usually rapid rather than gradual, because of internal thresholds within the climate system. This appears to have been as much the case with moisture levels at low latitudes as it was with temperature changes at high latitudes. Many tropical and subtropical areas also experienced abrupt, century-scale arid events during the course of the Holocene. Proxy data further suggest disequilibrium between earth surface processes (e.g. erosion, species distributions) and climate during previous climatic transitions, and this is likely to be true also of future, greenhouse-gas-induced warming.

Finally, the effects of future climate change on landscape stability will be complicated because their effects will be superimposed on existing human impact on vegetation and soils. As the case of dambo water resources illustrates, the resulting environmental conditions will result from a complex and subtle interplay between human and natural agencies. Human-induced land-use changes may also have an impact on climate at a regional scale. If the past is any guide, environmental change is not the exception, it is the norm.

NOTES

1   Carbon-14 (^{14}C) and uranium–thorium (U–Th) dating are methods of dating which rely on the radioactive decay of elements in the sample – the extent of decay indicating the age of the material.

2   The time-scales used here are based on calibrated radiocarbon ages; cal. yr BP = calibrated years before present.

REFERENCES

Bell, M. and Roberts, N. 1991. The political ecology of dambo soil and water resources in Zimbabwe. *Transactions of the Institute of British Geographers*, n.s. 16: 301–18.

Berger, A., Imbrie, J., Hays, J., Kukla, G., and Saltzman, B. (eds). 1984. *Milankovich and Climate*. NATO ASI Series C 126, 2 vols. Dordrecht: Reidel.

Bond, G. 1967. River valley morphology, stratigraphy and paleoclimatology in southern Africa. In Bishop, W. W., and Clark, J. D. (eds) *Background to Evolution in Africa*. Chicago: University of Chicago Press, pp. 303–12.

Bradley, R. 1984. *Quaternary Paleoclimatology*. London: Allen and Unwin.

Butzer, K. W. 1980. Adaption to global environmental change. *Professional Geographer* 32: 269–78.

COHMAP members. 1988. Major climatic changes of the last 18,000 years: observations and model simulations. *Science* 241: 1043–52.

Faulkner, R., and Lambert, R. 1991. The effect of irrigation on dambo hydrology: a case study. *Journal of Hydrology* 123: 147–61.

Gasse, F., and Van Campo, E. 1994. Abrupt post-glacial climate events in West Asia and North Africa monsoon domains. *Earth and Planetary Science Letters* 26: 435–56.

Gasse, F., Téhet, R., Durand, A., Gilbert, E., and Fontes, J.-C. 1990. The arid humid transition in the Sahara and the Sahel during the last deglaciation. *Nature* 346: 141–6.

Goudie, A. S. 1983. The arid earth. In Gardner, R., and Scoging, H. (eds) *Megageomorphology*. Oxford: Clarendon Press, 152–71.

Grove, A. T., and Warren, A. 1968. Quaternary landforms and climate on the south side of the Sahara. *Geographical Journal* 134: 194–208.

Knox, J. C. 1972. Valley alluviation in southwestern Wisconsin. *Annals of the Association of American Geographers* 62: 401–10.

Kutzbach, J. E., and Street-Perrott, F. A. 1985. Milankovich forcing of fluctuations in the level of tropical lakes from 18 to 0 kyr BP. *Nature* 317: 130–4.

Lamb, H. F., Grasse, F., Benkaddour, A., El-Hamouti, N., Van der Kaars, S., Perkins, W. T., Pearce, N. J., and Roberts, C. N. 1995. Relation between century-scale Holocene arid intervals in tropical and temperate zones. *Nature* 373: 134–7.

Lézine, A.-M. 1989. Late Quaternary vegetation and climate of the Sahel. *Quaternary Research* 32: 317–34.

Lorius, C., Jouzel, J., Raynaud, D., Hansen, J., and Le Treut, H. 1990. The ice-core record: climate sensitivity and future greenhouse warming. *Nature* 347: 139–45.

Lowe, J. J., and Walker, M. J. C. 1984. *Reconstructing Quaternary Environments*. London: Longman.

Mackel, R. 1985. Dambos and related landforms in Africa: an example of the ecological approach to tropical geomorphology. *Zeitschrift für Geomorphologie, Supplementband* NF 52: 1–23.

Mitchell, J. F. B. 1990. Greenhouse warming: is the Holocene a good analogue? *Journal of Climate* 3: 1177–92.

Mitchell, J. F. B., and Zeng Qingcun. 1991. Climate change prediction. In Jäger, J. and Ferguson, H. L. (eds) *Climate Change: Science, Impacts and Policy* (Proceedings of the Second World Climate Conference). Cambridge: Cambridge University Press, 59–70.

Nicholson, S. E. 1980. Saharan climates in historic times. In Williams, M. A. J., and Faure, H. (eds) *The Sahara and the Nile*. Rotterdam: A. A. Balkema 173–200.

Owen, R. B., Crossley, R., Johnson, T. C., Tweddle, D., Kornfield, I., Davison, S., Eccles, D. H., and Engstrom, D. E. 1990. Major low levels of Lake Malawi and their implications for speciation rates in cichlid fishes. *Proceedings of the Royal Society of London, Series B* 240: 519–53.

Petit-Maire, N., Fontugne, M., and Rouland, C. 1991. Atmospheric methane ratio and environmental changes in the Sahara and Sahel during the last 130 kyrs. *Palaeogeography, Palaeoclimatology, Palaecology* 86: 197–204.

Ritchie, J. C. 1994. Holocene pollen strata from Oyo, northwestern Sudan: problems of interpretation in a hyperarid environment. *Holocene* 4: 9–15.

Roberts, N. 1988. Dambos in development: management of a fragile ecological resource. *Journal of Biogeography* 15: 141–8.

Roberts, N. 1989. *The Holocene: An Environmental history.* Oxford: Blackwell.

Roberts, N., and Barker, P. 1993. Landscape stability and biogeomorphic response to past and future climatic shifts in inter-tropical Africa. In Thomas, D. S. G., and Allison, R. J. (eds) *Environmental Sensitivity.* Chichester: Wiley, 65–82.

Roberts, N., and Lambert, R. 1990. Degradation of dambo soils and peasant agriculture in Zimbabwe. In Boardman, J., Foster, I. D. L., and Dearing, J. A. (eds) *Soil Erosion from Agricultural Land.* Chichester: Wiley, 537–58.

Roberts, N., and Wright, H. E. 1993. Vegetational, lake level and climatic history of the Near East and Southwest Asia. In Wright, H. E., Jr, Kutzbach, J. E., Webb, T. III, Ruddiamn, W. F., Street-Perrott, F. A., and Bartlein, P. J. (eds) *Global Climates since the Last Glacial Maximum.* Minneapolis: University of Minnesota Press, 194–220.

Roberts, N., Taieb, M., Barker, P., Damnati, B., Icole, M., and Williamson, D. 1993. Timing of Younger Dryas climatic events in East Africa from lake-level changes. *Nature* 366: 146–8.

Roberts, N., Lamb, H., El-Hamouti, N., and Barker, P. 1994. Abrupt Holocene hydro-climatic events: palaeolimnological evidence from northwest Africa. In Millington, A. C., and Pye, K. (eds) *Effects of Environmental Change in Drylands.* Chichester: Wiley, 163–75.

Singh, D., Joshi, R. D., Chopa, S. K., and Singh, A. B. 1974. Quaternary history of vegetation and climate of the Rajasthan desert, India. *Philosophical Transactions of the Royal Society of London, Series B* 267: 467–501.

Street-Perrott, F. A., and Perrott, R. A. 1990. Abrupt climatic fluctuations in the tropics: the influence of Atlantic Ocean circulation. *Nature* 343: 607–12.

Street-Perrott, F. A., Mitchell, J. F. B., and Brunner, J. S. 1991. Milankovich and albedo forcing of the tropical monsoon: a comparison of geological evidence and numerical simulations for 9,000 yr BP. *Transactions of the Royal Society of Edinburgh, Earth Science* 81: 407–27.

Thomas, M. F., and Thorp, M. B. 1995. Geomorphic response to rapid climatic and hydrologic change during the late Pleistocene and early Holocene in the humid and subhumid tropics. *Quaternary Science Review* 14: 101–24.

Till, C., and Guiot, J. 1990. Reconstructions of precipitation in Morocco since 1110 A.D. based on *Cedrus atlantica* tree-ring widths. *Quaternary Research* 33: 337–51.

Verschuren, D. 1994. Sensitivity of tropical-African aquatic invertebrates to short-term trends in lake level and salinity: a paleolimnological test at Lake Oloidien, Kenya. *Journal of Paleolimnology* 10: 253–63.

Whitlow, R. W. 1985. Dambos in Zimbabwe: a review. *Zeitschrift für Geomorphologie, Supplementband* NF 52: 115–46.

3

THE CENTURY TIME-SCALE

Jean Grove

Editors' note Jean Grove's chapter looks at our understanding of environmental change over the past few millennia, and then concentrates on change in the past one millennium and its impact on particular human societies. She concentrates on the Medieval Warm Epoch (*c.* AD 900 to *c.* 1250–1300), during which the Vikings settled Greenland, and the Little Ice Age (*c.* fourteenth century to nineteenth century). The fluctuations in this period were so minor, especially in comparison with many that preceded them earlier in the Quaternary, that they might be expected to have been of little importance to humanity. But, as Grove shows, these fluctuations were important in human terms; they were changes with which different societies had to contend, some adapting successfully, others failing. Grove's illustration that in the more complex economy of medieval England during periods of stress there were gainers as well as losers (most of the latter being the poorest members of society) should be remembered when thinking how the contemporary world community may react to change.

INTRODUCTION

Instability is a leading characteristic of climate. Fluctuations take place on several different scales simultaneously, smaller disturbances being superimposed on larger. The smallest scale is what we call weather. This chapter is concerned with century-scale fluctuations, but before discussing their characteristics it may be helpful to place them in context.

During the past two million years or so, that is to say during the period in which mankind has separated in evolutionary terms from the higher apes, the globe has experienced a series of violent swings between glacial and interglacial conditions. Sediment cores from the deep oceans and ice cores extracted from the Antarctic and Greenland ice-sheets have revealed a sequence of oscillations with periodicities of about 100,000 years, long cold periods or glacials being separated by short warmer interglacials (Shackleton and Opdyke 1973). Even some of the less recent of these very harsh periods, in which great ice-sheets spread over large parts of Europe and North

39

America, were experienced by our ancestors. (It is worth noticing that had not at least some of them been able to adapt to changing circumstances *Homo sapiens* could not have survived.) During the Pleistocene, global average surface temperatures have varied by about 5–7°C, while in some middle and high latitudes temperature varied by as much as 10–15°C (IPCC 1990, p. 201). We know that the last glaciation and the interglacial which preceded it were themselves complicated by series of smaller oscillations. The glacial was made up of stadials and interstadials (Bond *et al.* 1993), and it appears from very recent investigations that the last interglacial was probably also a very complex period, subject to more violent climatic swings than the current interglacial, which would have affected the wild animals and plants, and so also the hunters and plant gatherers, especially, but not exclusively, those then occupying higher and middle latitudes (Lorius and Oeschger 1994). The inhabitants of low latitudes would also have had to respond to changing environment and especially to changes in moisture.

HOLOCENE FLUCTUATIONS

Climatic fluctuations since the last glacial – that is, during the past 10,000 years, the period known as the Holocene – have been very much less extreme. Global average temperature has probably varied through a range of about 2°C (IPCC 1990, p. 202), but the climatic environment has been far from uniform, either in low latitudes or in high latitudes. The warmest conditions occurred in the mid-Holocene. Consequently coniferous forests were able to extend farther north in both North America and Siberia between about 6000 and 4500 BP than is possible at the present day (Figure 3.1). Since the mid-Holocene irregular temperature decline has taken place. In high latitudes and in many of the high mountain regions of the world such as the Andes and Himalya, small, sensitive glaciers advanced a kilometre or two, fluctuated about their new positions, and then retreated again, several times during the Holocene.

It is likely that the changes in climate responsible for these expansions were world-wide. The only scientist who has carried out field investigations in Europe, Asia, New Zealand, and both American continents (Röthlisberger 1986) was convinced that the impact was global (Figure 3.2). It remains to be determined conclusively that all the periods identified were of global importance. There is, however, no doubt that the most recent one, known as the Little Ice Age (which lasted approximately from the fourteenth to the nineteenth century – but see below), did affect all continents (Grove 1988).

Small mountain glaciers react swiftly to changes in meteorological conditions, especially to summer temperature and winter precipitation. They leave traces in the form of moraines, accumulations of rock debris alongside their tongues or around their fronts. Many of these have organic material under, over, or within them which makes radiocarbon assessment of their

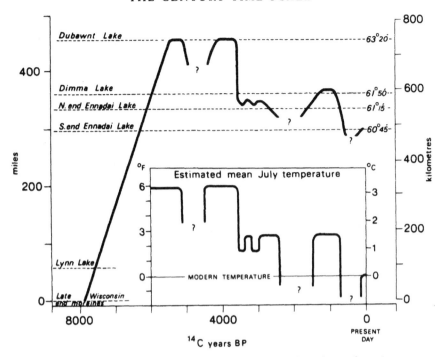

Figure 3.1 A reconstruction of the position of the northern limit of continuous forest along the 100°W meridian in central Canada during the Holocene, based mainly on radiocarbon-dated pollen diagrams from the Ennadai and Lynn Lake areas. The departures from the modern mean July temperatures were calculated from the varying distance of the forest limit from Ennadai. (After Nichols 1974, from Grove 1988)

ages possible. Changes in the extent and position of glacier tongues in Europe during recent centuries are often recorded in documents, as they affect farming and mountain communities. In some places the more recent histories are also recorded in sketches and pictures (Zumbühl 1980).

Glaciers provide us with proxy indications of changes in climate. This sort of data can be paralleled by others provided, for example, by sequences of tree-rings and the thickness, isotopic characteristics, and chemical and particulate characteristics of ice layers. It has to be remembered, however, that the various forms of proxy data have different sensitivities and response times to changes in climate. While tree-ring thicknesses relate to particular years and those immediately preceding them, mountain glaciers respond to the balance between accumulation and loss of snow and ice. Studies of the Rhône and other European glaciers have shown that the daily mean temperature for July and August accounts for 58 per cent of their variance, June precipitation for 16 per cent and total precipitation between October and May for 5 per cent (Reynaud 1980). That is to say that the pattern of their

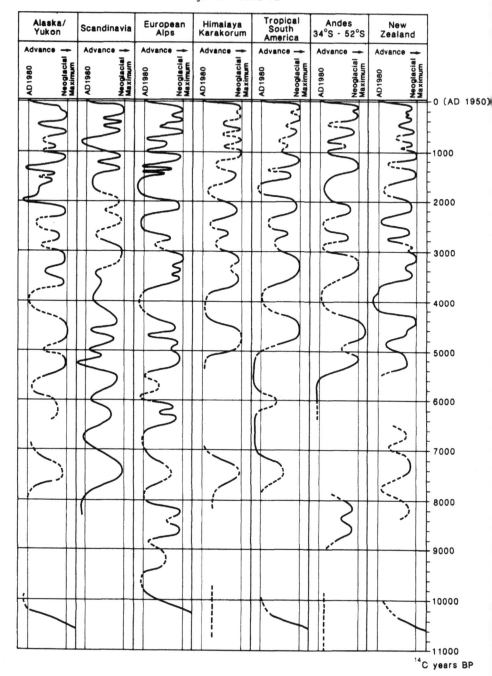

Figure 3.2 Fluctuations of glaciers in the Northern and Southern Hemispheres during the Holocene. (From Röthlisberger 1986, from Grove 1988)

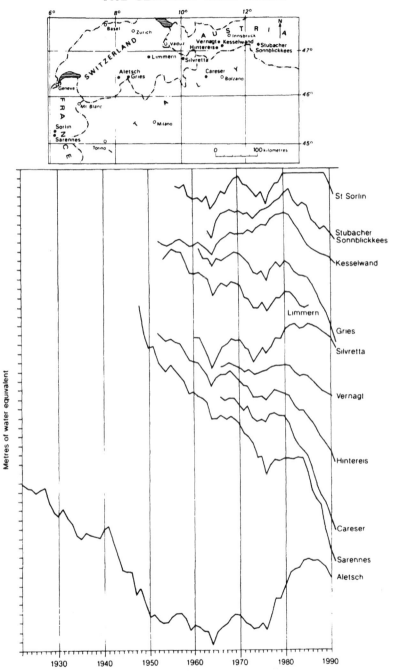

Figure 3.3 The cumulative mass balances of alpine glaciers.
(From Reynaud 1980, updated)

advance and retreat depends primarily on summer temperature and the amount of precipitation in June and over the winter and spring. If the annual increases or decreases of volume of the major Alpine glaciers made since measurements began are compared, we find that their trends of increase and decrease have been in concert (Figure 3.3). It is therefore clear that a history of the main advance and retreat phases of the Alpine glaciers provides an index of environmental fluctuations during the Holocene, which are also of obvious importance to the growth of plants and thus to agriculture and hence to human society.

GLACIAL HISTORY OF THE EUROPEAN HOLOCENE

The Swiss glaciers

The glacial history of the Holocene is best known for Europe, especially the Alps, partly because investigations of moraine sequences have been most intense here and partly because of the unusual wealth of historical data of many sorts (Figure 3.4). It emerges that periods of glacial extension lasting a few centuries have occurred at intervals all through the past 10,000 years. Some, such as the Lobben, between 3,000 and 3,600 BP included episodes which were rather more extensive than those of recent centuries. Many, and almost certainly all of them, were complex, depositing whole series of moraines, indicating that environmental conditions were variable within each such period; that is, that the century-scale fluctuation was itself composed of more minor oscillations.

During the past thousand years the basic pattern traced out by the glacier oscillations was threefold. Around AD 800 the large alpine glaciers advanced and remained enlarged for about a hundred years. Then followed the Medieval Warm Period, which lasted from about AD 900 until AD 1250–1300. Evidence for this is widespread outside Europe (Figure 3.5). Then the ice advanced once more at the beginning of the complex climatic period known as the Little Ice Age, which lasted to the late nineteenth century. Its general outlines can be represented by the fluctuations of the Grosser Aletsch Glacier (Figure 3.6). This glacier advanced almost as far in the century preceding the Medieval Warm Period as it did more recently. It is noteworthy that the warm period was itself broken by a minor advance centred around AD 1100. During the past few centuries there have been three advance phases culminating around 1350, 1650, and 1850. Between the first and second the ice retracted, but not to the extent it had during the Medieval Warm Period.[1] From this point of view we can see the Little Ice Age as being predominantly a period of glacial extension, broken by recession between 1450 and 1550, and with most prolonged extension between the late sixteenth and the mid-eighteenth centuries. Field evidence of the advances in the late thirteenth

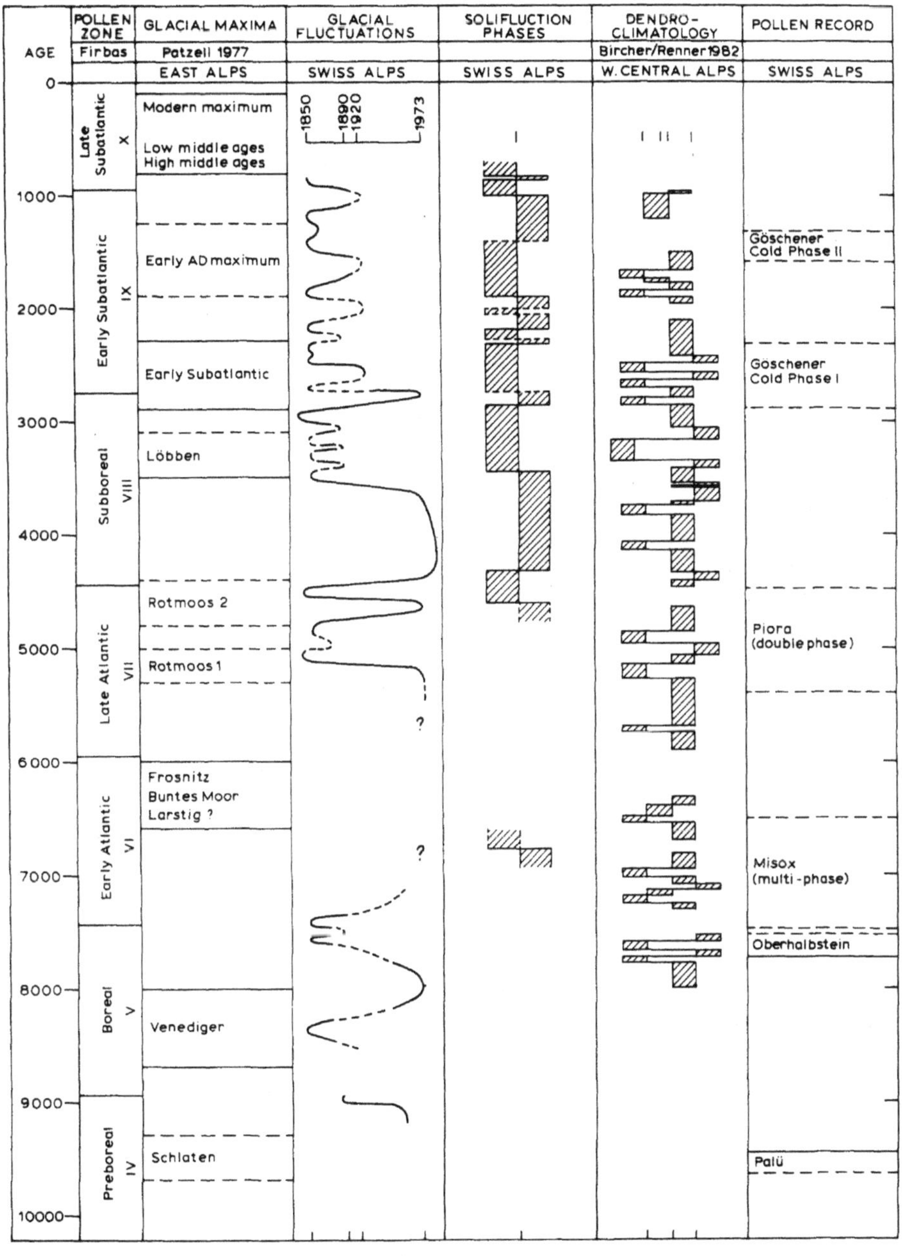

Figure 3.4 Environmental fluctuations in the European Alps during the Holocene.
As glaciers advanced and retreated, solifluction alternated with soil formation,
tree-rings narrowed and widened and plant species and their pollen varied.
(After Gamper and Switsur 1982, from Grove 1988)

45

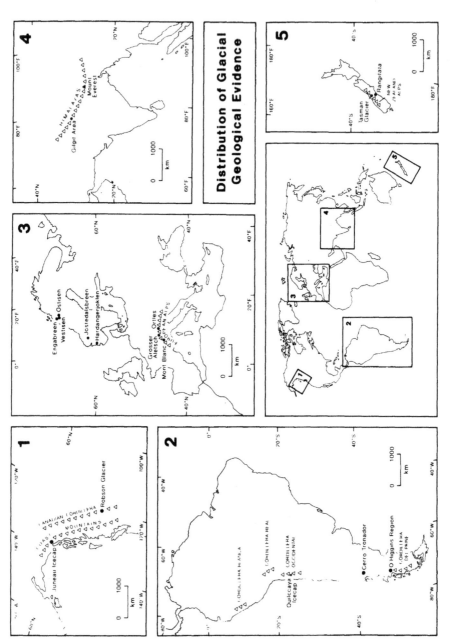

Figure 3.5 Radiocarbon dating of moraines indicates that glaciers withdrew during the Medieval Warm Period. (From Grove and Switsur 1994)

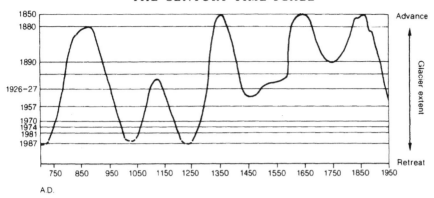

Figure 3.6 The fluctuations of the Grösser Aletsch Glacier. This diagram shows the positions of the front of the glacier since AD 750 compared with those it has occupied since AD 1850. (After Holzhauser 1984, from Grove and Switsur 1994)

century is preserved in locations where their traces have not been concealed by later deposits. This includes human artefacts such as the remains of the Oberrieden Bisse, a wooden irrigation canal constructed before being overridden by the Grösser Aletsch during the advances, and remnants of medieval roadways cut by the advancing ice (Röthlisberger 1974). It is also recorded in medieval documents (Lütschg 1926; Delibrias *et al.* 1975).

Not unexpectedly, data from the more recent past are much the most dense. They are particularly rich for the Swiss glaciers, partly because these were in full view of settlements, in a very literate country, with many early scientists and an astonishing number of artists. Also, deliberate monitoring of the Swiss glaciers began in 1880, earlier than in any other country, and has been continued consistently, providing a uniquely long and complete record. It has been possible to trace the history of the Lower Grindelwald Glacier since 1600 by using a combination of paintings, drawings, documents of several types, and finally the measurements made since 1880 (Figure 3.7).

Examination of the recorded snout positions of Swiss glaciers, and of others monitored elsewhere, reveals that fronts have typically exhibited small-scale variations in position in addition to the main phases of advance and retreat, already discussed (Figure 3.8). If the timing of advance, retreat, and stationary phases across all the monitored fronts of alpine Europe is considered, it is found that, despite minor variations from glacier to glacier, a clear pattern emerges (Figure 3.9). The recession of the past hundred years, like the advance phases which came before, was compound. Although all the glaciers are smaller than they were when measurement began, they experienced marked periods of increase in the 1890s, around 1920, and in the 1970s and 1980s. Decadal-scale variations are superimposed on the century-scale fluctuations during both advance and retreat. Since the late nineteenth century temperature has risen in the Northern Hemisphere by about 0.45°C,

47

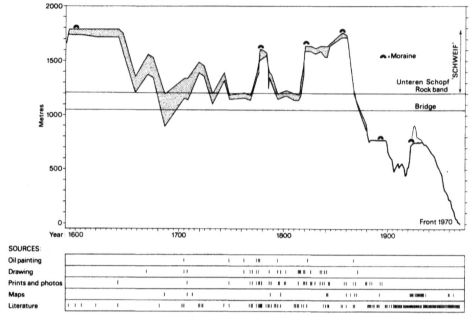

LOWER GRINDELWALD GLACIER 1590-1970

Figure 3.7 The frontal positions of the Lower Grindelwald Glacier, 1590–1970. The changing appearance and positions of the front of the glacier attracted so much attention from contemporaries that it is possible to trace its detailed history. (After Messerli *et al.* 1978, from Grove 1988)

but between the late 1930s and late 1960s temperature fell by about 0.2°C (IPCC 1990, pp. 206–7).

The meteorological controls of glacier fluctuations

The Swiss engineer Ignace Venetz was the first to use the behaviour of glaciers to show that 'temperature rises and falls periodically but in an irregular manner' (Venetz 1833). It is in Switzerland that the relations between minor glacier variations and their meteorological controls have been traced in most detail. Deliberate observation of temperature and pressure began in Basel in 1855, and other stations soon followed. A number of remarkably detailed weather diaries, such as that of Wolfgang Haller with records of frequency of sunny and rainy days near Zurich in 1545–6 and 1550–75, had been kept during the previous three centuries. Moreover, an extraordinary number of readily quantifiable items of information relating to weather, such as the dates of freezing of lakes, are to be found in Swiss archives. Christian Pfister (1981, 1992) collected a great volume of these, some 80,000 daily entries from diaries, about 33,000 phenological observa-

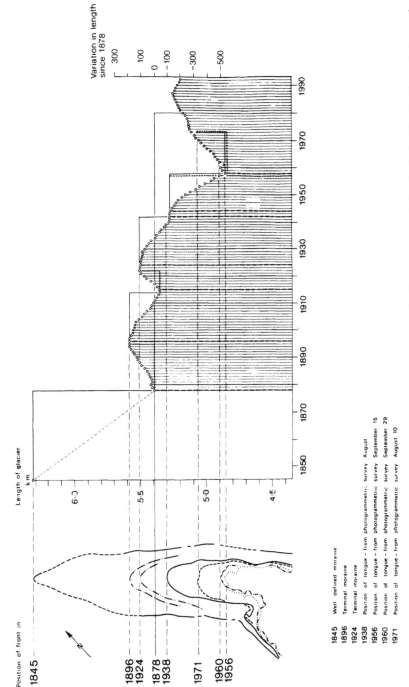

Figure 3.8 The changing extent of the Glacier de Trient, 1845–1990. Yearly observations made by the Swiss Glacier Commission make it possible to trace the minor oscillations of the tongue of the glacier. (From figures published in *Les variations des glaciers suisses*)

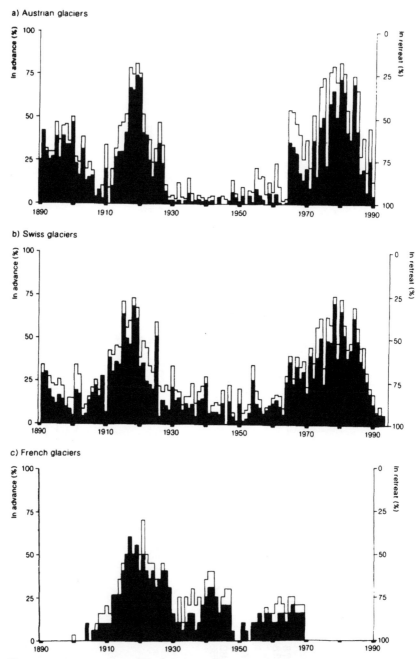

Figure 3.9 A comparison of the behaviour of glaciers in Austria, Switzerland, and France since 1890 in terms of the percentage of those observed each year which were found to be retreating, stationary or advancing. (From Grove 1988, updated)

tions, and hundreds of observations of snow cover at various altitudes, and used them to construct thermal and wetness indices, thus piecing together the climatic history of the sixteenth, seventeenth, and early eighteenth centuries. Phenophases – that is, the various stages of development of plants – can be related to monthly temperatures using multiple regression. The influence of altitude and exposure must naturally be taken into account.

Departures from mean values are the significant features of phenological records. For them to be used meaningfully it is, therefore, necessary to know whether the mean for the species has varied through time because of factors other than climate: the introduction of new varieties for instance. Pfister's computer-based methodology provides the means to quantify documentary evidence of past weather effectively, a necessary prerequisite to detailed investigation of its causes and effects (Pfister 1992).[2]

Pfister's temperature and precipitation indices, derived by computation from these data from the early sixteenth to the late twentieth centuries, are shown in Figure 3.10. The later part of the Little Ice Age in Switzerland was evidently made up of a series of minor fluctuations each lasting one or more decades. Variation was as great as in the twentieth century, but cold winters extending into March, and wet summers, with short growing seasons, were more prevalent than they have been in this century. Switches in the characteristics of individual months could be abrupt, so that even minor fluctuations were not usually homogeneous. The 1690s stand out as having been cold at all seasons, with heavy precipitation in spring, summer, and autumn. Cold winters and wet summers were more frequent than they have been since 1860 and these were often bunched together. Climatic reconstruction, like glacier history, indicates that a warmer period from the tenth to the twelfth centuries was followed by a period lasting into the nineteenth century which was colder (Jones 1990).

A rapidly increasing amount of information about the past climate in many other parts of the world has accumulated and is still doing so rapidly (e.g. Mikami 1992; Bradley and Jones 1992). Much of this is again based on documentary sources, on tree-ring sequences, and on the characteristics of ice-cores. It is therefore becoming increasingly possible to obtain an understanding of the complex spatial patterns of climatic change during the past few centuries on a hemispherical, if not a global, scale. It is evident that changes in the general circulation of the atmosphere have resulted in positive anomalies in some areas at the same time as negative anomalies in others. This is not at all surprising in view of the patterns disclosed by the instrumental records obtained during the current century (Briffa and Jones 1993). No one would deny that this has been a time of global warming, yet there was an unexpected drop in average temperature in the Northern Hemisphere of about 0.2°C after about 1940 associated with an increase of the region covered with lying snow in winter in Eurasia and the marked increase in the proportion of advancing glaciers in the Alps (Figure 3.9). The results brought

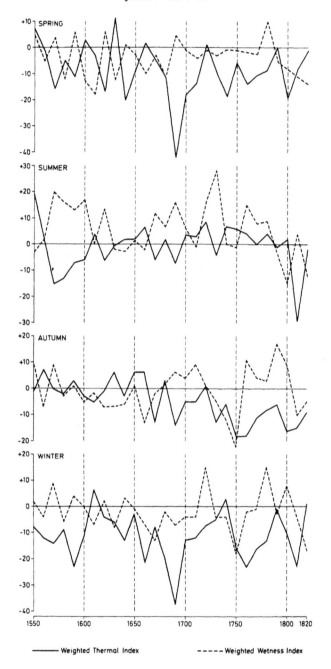

Figure 3.10 Seasonal weighted thermal and wetness indices for Switzerland, 1550–1820. (Based on data in Pfister 1981, from Grove 1988)

together by Bradley and Jones from a great number of sources clearly demonstrate that the Little Ice Age was not a long, synchronous cold period. This is also not surprising in view of the complex fluctuations exhibited by the glacial record. Yet the glacial history of the Medieval Warm Period and Little Ice Age clearly indicates that the small glaciers on all continents advanced and retreated in general concert with each other, though there were some relatively minor regional variations in timing.

The explanation of the apparent discrepancy can be seen as due to three main factors. First, when Bradley and Jones expressed the view that 'the last five hundred years was a period of complex climatic anomalies.... There is no evidence for a world-wide synchronous cold interval to which we can ascribe the term "Little Ice Age"' (1992), they had not then attempted to

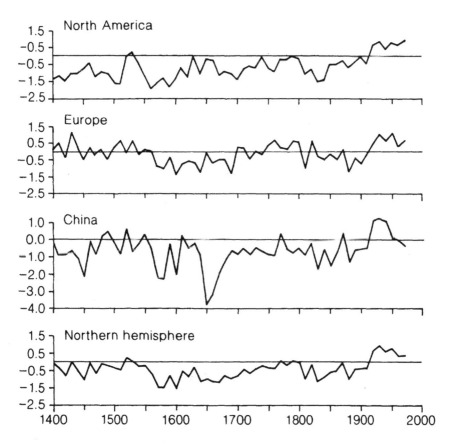

Figure 3.11 Composite temperature anomaly series for Europe, North America, East Asia, and the Northern Hemisphere derived from normalised decadally averaged anomalies, with reference to the period 1860–1959.
(From Bradley and Jones 1993)

combine the various series to give regional or hemispherical values. When they came to do so (1993), they produced a composite temperature anomaly series for the Northern Hemisphere which was very much what might have been expected from the glacial evidence (Figure 3.11), which they had been disposed to dismiss. Second, the complexity of many moraine sequences is in any case suggestive of sequences of small-scale climatic fluctuations rather than unbroken cold phases. Third, their data concerned summer temperatures. But changes in glacier mass relate to precipitation as well as ablation, to winter as well as summer conditions. It is instructive to notice that the small-scale depression of temperature in recent decades was clearly reflected in the pattern of advance and retreat of Alpine glaciers (Figure 3.9). It is also relevant to notice the accordance of timing between glacier advances and the occurrence of phenomena such as landslides and avalanches (Grove and Battagel 1983).

THE CONSEQUENCES OF CENTURY-SCALE AND DECADAL-SCALE ENVIRONMENTAL CHANGE: CASE STUDIES

Although it is generally agreed that the temperature anomalies associated with the occurrence of the Medieval Warm Period and Little Ice Age were no more than 1 or 2°C, the physical and biological consequences of both were marked, and many of them of such a nature as to have direct human implications. From the point of view of human time-scales a number of questions seem to arise. First, have any of the main climatic periods of the Holocene had obvious important historical consequences? Second, are such consequences limited to high latitude, and high-altitude marginal areas? Has the impact of closely spaced sequences of years marked by very low or high temperatures, very benign or very difficult conditions, been of historical importance? Lastly, has the history of past climatic fluctuations to be taken into account in forecasting the probability of future climatic scenarios?

The Medieval Warm Period and its decline

Greenland

It is clear that Vikings could not have colonised Greenland in the manner that they did had it not been for the relatively benign climatic circumstances of the Medieval Warm Period. When Erik the Red and his companions set out for the west coast of south Greenland in 986 it was to a land which Erik had already investigated, and he knew it offered grassy valleys and fjord sides suitable for livestock (Gad 1970). The intention was to settle and farm in the traditional Viking manner, based on animal husbandry; and the colonists may even have attempted to grow cereals. The whole enterprise was basically

54

dependent upon the grass crop, together with fishing and sealing carried out with such boats as the group was already accustomed to use. There was evidently no intention of changing the basic way of life. The two Greenland settlements, the Western Settlement, in what is now the Godthåb district, and the larger Eastern Settlement, in the modern Narssaq and Julianehåb District (Figure 3.12), survived much as Erik might have expected for at least a century and a half before deterioration set in. Goods such as honey, needed to make mead, were no doubt much-valued imports made possible by the infrequent visits by small ships from Scandinavia (Gad 1970). But basic food supplies were home procured.

Success depended on skilful coordination of communal labour to make maximum use of seasonal abundance of both terrestrial and marine resources (McGovern 1981; Figure 3.13). Livestock had to be kept inside during winter, while the settlers survived on dried meats and dairy products. Excavations have revealed that even the farms farthest inland were basically dependent upon seal meat, particularly that of the harp seal (*Pagophilus groenlandicus*). Much of the population was probably involved in sealing during the annual migration of these animals up the west coast in May–June, in harvesting the hay in summer, and then in caribou hunting in autumn. On top of all this there were yearly hunting trips to Nordseter, the hunting grounds over 800 k to the north, round Disko Bay. They provided the walrus tusks and bearskins used for such trading as was possible. McGovern (1981) argued persuasively that the economy always operated on such a thin edge that slight shifts in the extent and timing of environmental fluctuations would have had serious repercussions.

Evidence of warmer conditions during the Medieval Warm Period comes not only from the very existence of the settlements, but also from the independent evidence from ice-cores (Dansgaard *et al.* 1975). There is also independent evidence of the deterioration of climate which eventually led to the disappearance of the Viking settlements. The Western Settlement, with its ninety farms and four churches, was abandoned in mysterious circumstances, which left a few domestic animals wandering untended, sometime between 1341 and 1362, cutting off access to Nordseter. The Eastern Settlement declined more gradually, lost outside contact, and had died out altogether by AD 1500.

During the period 1954–75 there was minor cooling and increased interannual variability. The impact on the harp seal catch has been documented: the percentage of harp seals in the total catch at the inner fjord station of Kapisigdlit fell from 30 per cent in 1954–8 to 4 per cent in 1959–74. Harp seals are migratory, arriving in west Greenland in spring and moving up the west coast in large numbers, most remaining in the outer fjord zone. No harpoons or barbed spears have been uncovered during the many excavations of the Norse settlements, so the seals must have been taken in communal boat drives which forced them on to beaches or into nets. If the

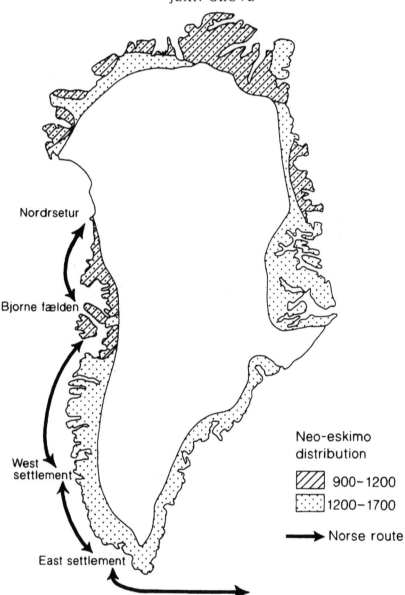

Figure 3.12 Distribution of the Norse settlements in Greenland. (After Gad, 1970)

minor cooling of the late twentieth century was sufficient to have such a marked effect on the migration of the harp seal, then it is reasonable to assume that the more prolonged and severe late-medieval cooling would have had even more pronounced consequences. At the same time lengthening of

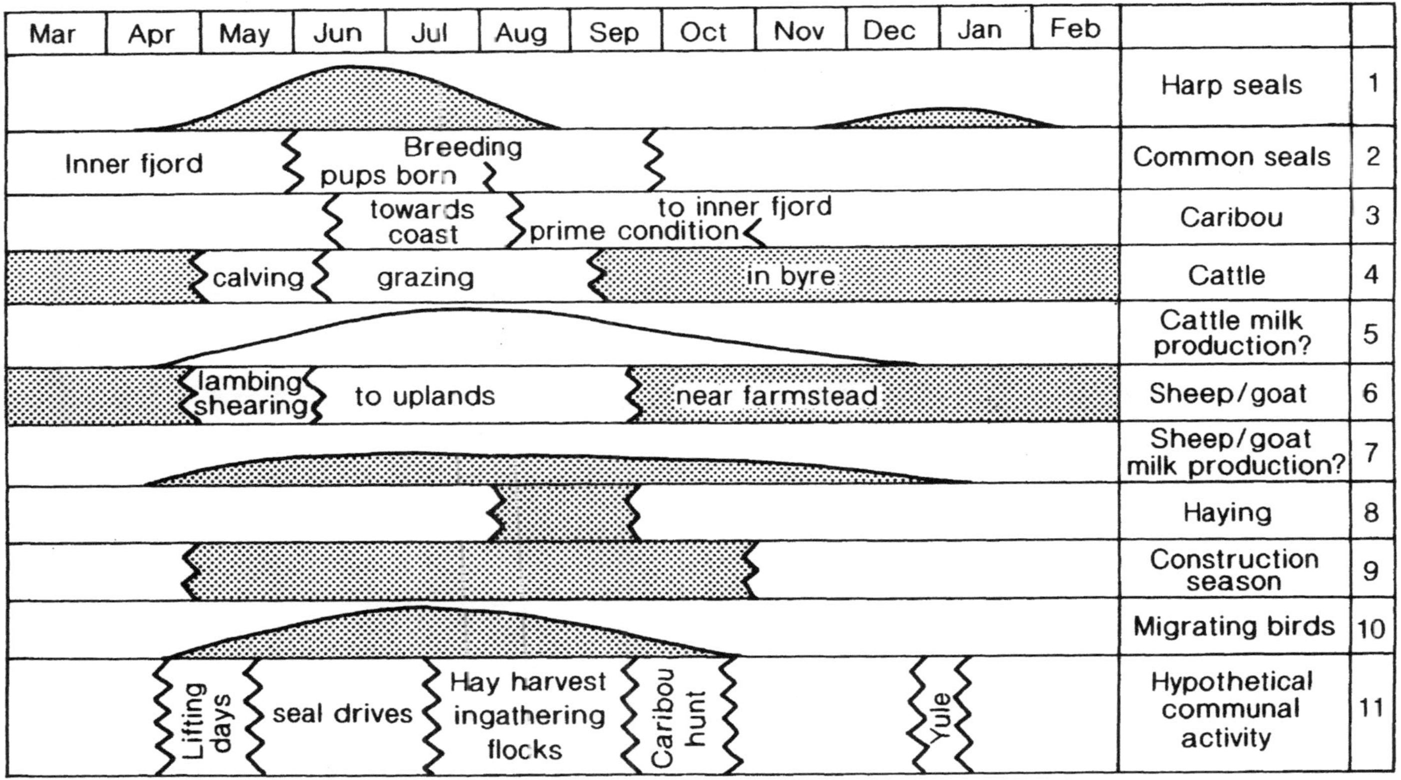

		Mar	Apr	May	Jun	Jul	Aug	Sep	Oct	Nov	Dec	Jan	Feb
Harp seals	1												
Common seals	2	Inner fjord			Breeding pups born								
Caribou	3					towards coast	prime condition	to inner fjord					
Cattle	4			calving	grazing			in byre					
Cattle milk production?	5												
Sheep/goat	6			lambing shearing	to uplands			near farmstead					
Sheep/goat milk production?	7												
Haying	8												
Construction season	9												
Migrating birds	10												
Hypothetical communal activity	11		Lifting days	seal drives		Hay harvest ingathering flocks			Caribou hunt		Yule		

Figure 3.13 The hypothetical Norse seasonal round. Note the concentration of activity in the few summer months.
(After McGovern 1981)

the winter season and greater persistence of winter snow cover would have increased the byring time of the domestic animals and simultaneously reduced the hay harvest. Furthermore, deep winter snow, especially if associated with crusting, is known to have caused dramatic reduction and even local extinction of caribou herds in south-west Greenland. The Norse settlements certainly suffered severe climatic stress and a reduced resource base after the first century and a half, which caused food supplies to become scarce.

Excavations have revealed not only the effects of malnutrition, but also a fall in soil temperature. Remains exhumed from early graves, now within permafrost layers, in the churchyard at Herfojolfsnes, in the East Settlement, were completely decomposed. The climate must have deteriorated sometime after burial, and decomposition had taken place.

Southwestern Colorado

The Anasazi people of the Dolores areas of southwestern Colorado (Figure 3.14) unlike the Greenlanders, exhibited a strong capacity to adapt to changing circumstances (Schlanger 1988). The prehistoric cultures of the American Southwest have attracted a great deal of attention from archae-ologists, among whom the general consensus is that there has been a striking correspondence between large-scale regional population trends and long-term climatic fluctuations on the southern Colorado Plateaus (e.g. Euler *et al.* 1979; Slatter 1979). Population rose in areas and periods offering favourable conditions for agriculture based on maize but declined in those areas and times unfavourable for such agriculture. In the Dolores area, in particular, the Anasazi adopted a strategy of population movement and relocation to support substantial population growth between AD 600 and 1250. In addition, after about AD 1100, they adopted agricultural intensifica-tion measures in response to local geographical and environmental circum-stances. Their horticultural economy was based on corn, beans, and squash. The Anasazi area in the southwestern corner of Colorado is rather simple topographically, but sensitive to small variations in climate affecting agricul-tural success.

The Great Sage Plain of southwest Colorado is cut by a dendritic canyon system running from northeast to southwest (Figure 3.14a). The dominant vegetation is sagebrush, with pinyon and juniper growing along ridges, more thickly in the upper reaches than the lower parts of the region. The canyon systems, such as the McElmo, house riparian sections with cottonwood and willow along the watercourses. Free water is confined to the canyon systems; otherwise the region is directly dependent on rain for watering crops. A transect from northwest to southeast provides representative samples of the current range of environments within the area. The Dolores Archaeological Program study area in the north has an annual average rainfall of 460mm, but a growing season close to the minimum of 110 to 130 days necessary for the maturation of aboriginal varieties of corn. The Dolores Canyon rim at 2,320m currently has a growing season of about 130 days, though the

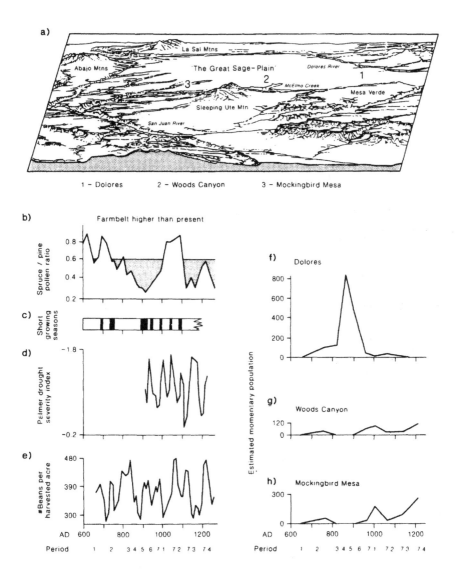

Figure 3.14 Anasazi adjustments to climatic fluctuations in southwestern Colorado. (a) The regional setting; (b), (c), (d) Patterns of long- and short-term environmental fluctuation; (e), (f), (g) Estimated population levels in Dolores, Woods Canyon and Mockingbird Mesa. (After Schlanger 1988)

Table 3.1 Time periods defined by Anasazi ceramic assemblages

Period	Time span (AD)
1	600–720
2	720–800
3	800–840
4	840–880
5	880–920
6	920–980
7.1	980–1025
7.2	1025–1100
7.3	1100–1175
7.4	1175–1250

growing season within the canyon may be shorter owing to cold air drainage and in some years may be less than 100 days. In the south Mockingbird Mesa, ranging from 1,830 to 1,950m in height, received close to 189mm rain annually. The growing season at the lower end of the Mesa is about 145 days.

At the present day dry farming, producing fair yields of such crops as beans, corn, wheat, and alfalfa, is restricted to altitudes between 2,010 and 2,320m which are high enough to receive ample rainfall and low enough to avoid early frosts.

A tree-ring-dated ceramic chronology has been developed which has proved useful in identifying and dating both longer- and shorter-term environmental changes which affected Anasazi culture (Table 3.1). Long-term changes have been modelled by Petersen (1988) using changes in the ratio of spruce (*Picea engelmanni*) to pine (*Pinus* spp.) pollen and changes in pollen influx frequencies of pinyon pine (*Pinus edulis*) in the pollen of Beef Pasture, a site at 3,060m 30km from Dolores. He interpreted a decline in spruce relative to pine as a response to decreases in winter precipitation and increases in pinyon pine influx as a response to increased summer precipitation. These pollen records are then used to establish the location of a 'farmbelt' zone with adequate rainfall and growing season length (Figure 3.14b). (The shading in this figure indicates periods when the lower boundary of the farmbelt was higher than at present.)

It appears that during AD 600–800 (periods 1 and 2) most of the area except the southern third fell within the farmbelt. But from 800 to about 1000 (period 3 to the middle of period 7.1) and again from 1100–1250 (periods 7.3 and 7.4) the lower two-thirds of the area fell outside the farmbelt, so that potential arable land was restricted to the northeast of the transect, whereas from about 1000 to 1100 (the middle period 7.1 to the end of 7.2) all the area fell within the farmbelt.

Despite these environmental shifts the Anasazi population grew substantially over the whole period from 600 to 1250. This seems to have been possible because they shifted their farming operations from altitude to altitude up and down the transect in order to solve the difficulties of low rainfall and short growing seasons. However, they were faced not only by long-term but also by short-term climatic fluctuations. Analyses of high-frequency fluctuations have been based on retrodiction of drought severity and of potential prehistoric yields of dry bean and corn using tree-ring sequences from several sites in southwestern Colorado and the Colorado Plateau (Figures 3.14c and 3.14d). Particularly interesting was Burns's (1983) retrodiction[3] of regional crop yields of dry beans and corn. Retrodicted harvests in excess of the long-term mean were taken to have produced surpluses and those below the long-term mean to have produced shortfalls (Figure 3.14d). Potential shortfalls seven to twenty-four years long were identified between AD 707 and 727, 751 and 760, 870 and 879, 995 and 1003, 1035 and 1041, 1146 and 1155, 1170 and 1193, and 1240 and 1246. These shortfalls would have been more severe at lower altitudes than higher, as the reconstruction was based more on precipitation than growing-season length. No lengthy shortfalls were indicated between 760 and 870, 880 and 995, or 1042 and 1146.

It is surmised by analogy with known buffering and short-term relief measures documented in the ethnographic southwest that responses to short-term difficulties would have included the use of fields in a variety of topographic settings, cultivation of several varieties of corn, beans, and squash with different drought-resistant qualities, corn storage to meet shortfalls, and redistribution of stores and supplies. Settlements would probably be abandoned after more prolonged droughts with durations of the order of eight years or more of combined drought and crop failure (Slatter 1979). Poor-quality diets, especially when wild famine foods were depleted, would have been especially harmful to the very young. Significant population reduction would probably have followed runs of as much as five years of low crop yields and absolute crop failures.

The combination of long- and short-term environmental fluctuations indicates that circumstances were not particularly favourable for corn and bean agriculture across the Dolores region during much of the period between AD 600 and AD 1250. Yet this subsistence population as a whole grew, the size and number of their settlements increasing. Population growth was quite rapid during 600–880 (periods 1–4, Figure 3.14e) and especially fast during 800–80, when both short- and long-term conditions were most favourable for dry and run-off farming (immigration, perhaps from low-lying areas, may have been involved). The short-term disturbances at the end of period 1 and again at the end of period 2 seem to have had less effect than long-term conditions. Despite continued favourable long-term conditions, especially at higher altitudes, population levels from AD 880 to AD 1025 (period 5 to the end of 7.1) fell in the Dolores area at the top of the study transect. Meanwhile population increases at lower elevations began in 920–80 (period 6), despite poor long-term conditions there, and continued during 880–1025, by which time long-term conditions would have been conducive to agricultural success. It seems that after 880–920 (period 5) the history of the Anasazi population and its distribution had less to do with long-term climate fluctuations than with the nature of the short-term fluctuations. After that time critically short growing seasons plagued the upper altitudes frequently, and during 1175–1250 (period 7.4) long-term conditions favoured agriculture at lower levels (Figure 3.14g).

The two principal hazards faced by the Anasazi farmers were short growing seasons and drought. Unfortunately, locations selected to meet water shortage were likely to run into the problem of inadequate growing seasons and early frost (Cordell 1984). The prehistoric farmers could do nothing to mitigate early frosts except move to lower altitudes, but they could try to alleviate drought by watering. The strategy of population movement and relocation seems to have been adequate to maintain and even increase the population between 600 and 1100. After 1100, by which time the total population was enlarged, the combination of adverse short-term conditions at higher altitudes and adverse long-term conditions at lower altitudes made it impossible to solve the consequent agricultural difficulties by moving upslope. This situation was met by the adoption of intensified techniques including the construction of reservoirs, tanks, cisterns, and irrigation ditches, which were introduced about AD 1100.

Medieval Britain

Southeast Scotland

In southeastern Scotland, as in much of northwestern Europe, higher-altitude arable land has commonly been used to grow oats. Oats above 250 m in Britain are sensitive to summer warmth and wetness and to exposure, which may be measured in terms of accumulated temperature, average wind speed and end-of-summer potential water surplus (Parry 1975). In the Lammermuir Hills minimum levels of summer warmth for the ripening of oats have been established to be 1050 degree-days above a base of 4.4 °C, and the maximum tolerable levels of the other two limiting factors are 6.3 m/s average wind speed and 60 mm potential water surplus. The zone which is climatically marginal for the harvesting of ripened oats can be identified by contouring the combined isopleths of these three parameters (Figure 3.15).

It has been estimated that if temperature over the period 1250–1450 fell by a little less than 1 °C, then summer warmth at 300 m in northern England would have been reduced by 15 per cent, the frequency of crop failure increased sevenfold from 1 year in 20 to 1 year in 3 and the frequency of crop failure two years running increased 70 times (Parry 1975, 1976, 1978). These estimates clearly indicate that there must have been substantial decreases in

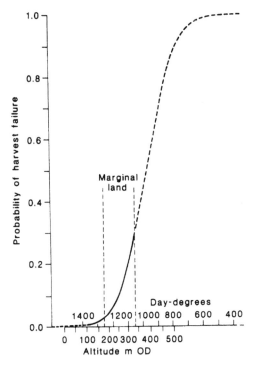

Figure 3.15 The probability of harvest failure in southeast Scotland. (From Parry 1976, from Grove 1988)

the viability of upland farming in both lowland Scotland and upland England in late-medieval times. That land abandonment took place in much the way that these calculations suggest was demonstrated by Parry (1975) and is shown in Figure 3.16. Upland in other parts of Britain was abandoned at much the same time as in the Lammermuirs; on Dartmoor all settlements above 300 m had been deserted (Beresford 1981).

Lowland England

It is generally accepted among historians that in the eleventh and twelfth centuries the population of England increased markedly. In the farming district of Taunton, belonging to the bishop of Winchester, the taxable male population over 12 years old rose from 612 in 1290 to 1,448 in 1311 (Titow 1972; Figure 3.17). This trend was not in fact restricted to England, although elsewhere the rapidity of growth might not have been quite so high. According to Abel (1980), the population in the Moselle region trebled between 1000 and 1237 and that of Germany grew slowly at the turn of the

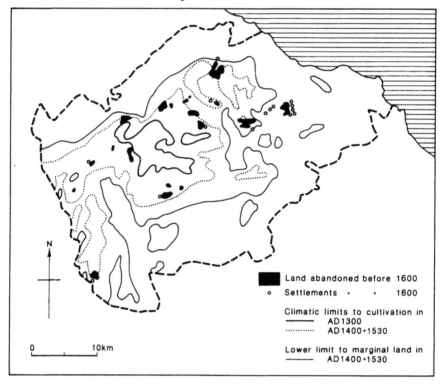

Figure 3.16 Abandoned land and lowered limits to cultivation in southeast Scotland between AD 1300 and 1600. (From Parry 1975, from Grove 1988)

millennium, gaining speed rapidly during the twelfth century to reach a peak in the thirteenth century.

Similar trends occurred in other parts of Europe, both northern and southern. Population increase may be attributed to a range of factors including extension of cultivated land,[4] increasing food supplies, improvements in agricultural and associated techniques, such as the change from two-course to three-course rotation, the introduction of the two-piece flail, and the spread of water-mills. Changes in social and economic organisation and the rise of towns no doubt also played their part. It is striking, however, that most historians have failed to discuss the relative absence of harvest failure during the Medieval Warm Period.[5] This omission is brought into high relief by attention paid to the sequence of harvest failures, many of them disastrous and associated with widespread famines which took place in the fourteenth century right across Europe (Lucas 1962[6]).

Many historians have been disposed to doubt whether the harvest failures of the thirteenth and fourteenth centuries were part of a more general climatic deterioration. It is argued here that the sequence of failures was in

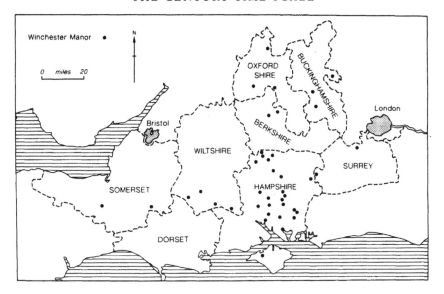

Figure 3.17 The distribution of the Winchester manors. (From Titow 1972)

fact associated with just such a deterioration, but that the longer-term climatic oscillation involved was itself constituted by an increased incidence of decades marked by excessive summer wetness or dryness, or by colder conditions (Farmer and Wigley 1983).

Particular attention will be focused here on a series of agricultural catastrophes experienced in England between 1315 and 1322 (Kershaw 1976), exactly during the period when the Swiss glaciers were advancing particularly swiftly (Figure 3.6). The crops of 1314 had been harvested only with great difficulty because of wet conditions and was deficient, so that many people were already in serious difficulties before the famine which came in 1315–16. The archbishop of Canterbury ordered processions and services of supplication in the summer of 1315. The harvest that year failed because of torrential rain and widespread flooding, which ruined both hay and cereal crops. England was already suffering from inflation, attributed to the importation of over-much silver gained by wool exports (Harvey 1976), but the rise in grain prices between 1315 and the following year from 8s. per quarter to 26s. 8d. was certainly the direct result of the weather. The many grants of protection and safe-conduct made to merchants in the winter and spring of 1316 indicate the profits which were to be made from long-distance grain trading. Not surprisingly the dearth was not felt to the same extent everywhere: food was exported from Cornwall, relatively little affected, to London. The price of salt rose as much as that of wheat, because of the lack of sun for drying. All types of food were in short supply.

By the spring and early summer of 1316 a major famine had resulted, and this was accompanied by a virulent epidemic, possibly of typhoid, which increased mortality. There is insufficient evidence to make a satisfactory estimate of the death rate in the country as a whole, but it is significant that the heriots paid on death in a group of Winchester manors suggest a crude rate of 10 per cent and the many accounts of unburied bodies littering city streets indicate that the rate was much higher in towns.[7] The suffering, which caused a very marked rise in crimes relating to theft of food, was increased by the decrease in alms-giving by rich households and religious houses.

The harvest of 1316 was even worse than that of the previous year. According to Kershaw, the two years 1315–16 and 1316–17 caused an inflation in grain prices unparalleled in English history. The harvest of 1317 was better and that of 1318 good, and prices fell lower than in any year since 1288. The worst of the famine seems to have been over by then, but not the end of the agrarian crises. The famine had been accompanied by widespread sheep murrain,[8] with mortality especially high among lambs and yearlings. Sheep grazing heavy soils and flooded grasslands were especially vulnerable. The effects varied greatly from place to place, but had largely passed by 1317. But in 1319 an epidemic of disease among cattle and oxen, probably imported from France, began and caused widespread destruction of herds through the period 1320–21. Then came a mediocre harvest in the wet summer of 1320, followed by a disastrous one in 1321, which was associated with drought in some areas (Farmer and Wigley 1983). In southeast England a serious drought, coupled with coastal flooding[9] and another major livestock epidemic in 1325 and 1326, had grave consequences. Christ Church in Canterbury claimed in 1327 to have lost 1,212 acres to the sea, 257 oxen, 522 cows and their issue, and 4,585 sheep.

That this sequence of largely weather-related disasters had major effects on both the economic and social fabric of the country cannot be denied. It is equally clear that the impact varied from place to place and class to class. Low-lying ground and valley bottoms suffered more than uplands with good drainage, so long as the upland farms were not too high. The effects of livestock disease not only had direct effects on food supply, but also caused serious shortages of manure, vital for arable production, and caused exports of wool to drop by just below one-third in the 1215/16–1334/5 period. Cattle disease[10] also led to an acute shortage of draught animals, in some places so great that the land could not be tilled. It was twenty years before stock levels recovered.

The poor suffered most, though even the rich were not immune. On many large estates the famine years brought a downturn in demesne production and exploitation. The capital required to replace livestock had to be found. In some places labour was short. Some arable reverted to pasture, some pasture to waste. For the fortunate few, with a surplus to sell, the crisis brought profits. But for others it was followed by a lengthy depression, whether of grain or wool production, which ended or hastened the end of demesne economy. Meanwhile there is abundant evidence that many smallholders were obliged to sell out, thereby allowing some few of their better-off neighbours to acquire more substantial holdings, thus further changing the social fabric and furthering the disintegration of the traditional tenurial structure. In the most severely hit regions considerable acreages went out of cultivation. This played its part in the desertion of villages, which was thus well under way before the Black Death[11] came.

The most detailed information about the effect of these hard years on yields comes from the estates of the bishopric of Winchester (Titow 1972)[12] (Figure 3.17). Titow, following his very detailed analysis of all the available data for 1209–1349, found that there had been a general deterioration in production towards the end of the thirteenth century, followed in many cases by recovery during the first half of the fourteenth century. On closer examination, however, he found that only ten to twelve cases of fourteenth-century recovery seem to have been instances of real improvement in fertility against some thirty to thirty-four which could be explained by contraction of the arable alone; that is, they were due, he considered, to contraction of the demesne arable to the most fertile parts of the total areas. Moreover, his graphs of annual average yields for the Winchester estate show violent interannual variations, which are particularly marked in the late thirteenth and early fourteenth centuries (Figure 3.18). There do not appear from these graphs to have been anything like sustained increases in yield between 1300 and 1340.

Titow considered that it was 'quite impossible to isolate and even more so to quantify the effects of changing weather conditions on the level of productivity' (1972, p. 24). He believed that the references to weather found in the account rolls, which were incidental to the purposes of the documents, did not provide an adequate guide to the vicissitudes of climate, although he went to the trouble of extracting and publishing them in full, being aware of their basic veracity (Titow 1959–60). He argued that if years with good or bad yields were in fact due to climate, then both good and bad occurred over the period, which he thought did not provide support for progressive climatic deterioration. At the time he wrote, the complexity of climatic fluctuations was not generally recognised among historians.[13] He was disposed, rather perversely in view of his detailed examination of the weather data, to the view that it was soil exhaustion, following sustained increases in population, which was the root cause of yield deterioration. He also believed, very reasonably, that shortage of manure would have been a crucial factor, the ratio of animals to area under seed having fallen catastrophically because of the various murrains. He noted the strength of the hypothesis of a strong causal link between falling fertility and previous expansion of the arable. As arable cultivation contracted it was in those manors where assarting and colonisation of the waste had been greatest in the thirteenth century that the greatest deterioration of yields took place. Finally, he acknowledged that as

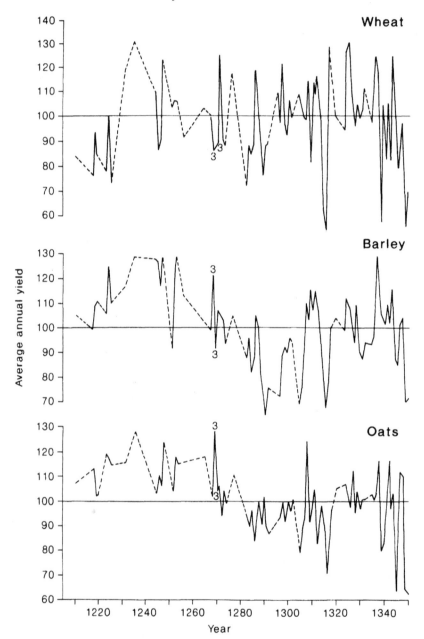

Figure 3.18 Average annual yields for the Winchester estates. Each year's average is calculated against the 1209–1350 average of the manors going into the annual average. 3 = average of only 3 manors. (From Titow 1972)

a considerable deterioration in yields also occurred on large manors which did not seem to have colonised to any great extent in the previous century, over-cultivation was obviously not the sole answer.

The fluctuations in yield shown in Figure 3.18 are not the type which would be expected from the effects of serious soil exhaustion, further exaggerated by lack of manure. It is very noticeable that the Norwich Climatic Research Group's re-examination of the documentary data from the period 1220–1439 (Table 3.2), which made use of the Winchester data corroborated by verified data from other sources relating to southeast and central southern England, shows lack of precipitation in May, June, July, October, November, and December of 1310, May and June 1311, April, May and June 1312, June and July 1313. Yields in these years were presumably affected by drought. The extreme wetness from August 1313 until February 1317 emerges clearly, substantiating Kershaw's descriptions. Conditions were made harder by unusually low temperatures in the winters of 1314, 1315, and 1317. There are no indications of either heavy precipitation or drought in the summer and autumn of 1317, when the harvest was somewhat better. The good harvest of 1318 was no doubt promoted by the dry conditions in June and July. The months of August to November 1320 were very wet, accounting for the mediocre harvest. While April and May 1321 were wet, June and July were dry, confirming that the disastrous harvest that year was indeed associated with drought in some regions. The results of the Norwich study produce strong confirmatory evidence of the weather sequence suggested by Kershaw. The decadal indices for 1220 to 1450 produced by the Norwich Group are shown in Figure 3.19, indicating substantial variability in both temperature and moisture and a notable downturn in winter temperatures in the latter part of the period.

Titow (1972, p. 24) rightly concerned about the danger of a circular argument 'whereby years of good and bad weather are defined and severity of conditions measured, in terms of yield calculations and the information so obtained is then used to explain the yields themselves'. This difficulty is completely avoided by the use of the Norwich data, which are obtained quite independently of considerations of yields, but none the less explain Titow's results in surprising detail. It is not suggested here that none of the other factors identified by Titow played a part in reducing yields. Lack of manure in particular must have been important. It is suggested, however, that the importance of soil exhaustion, suggested by Postan (1975), has been overplayed.[14] Population expansion, promoted at least in part by periods without harvest failure, would certainly have led to expansion into marginal areas with less fertile soils. It is, however, surely much more likely that it was bad weather conditions which led to decreasing yields, which would have been felt first in more marginal areas, rather than soil exhaustion, for which we can have little direct evidence. The sequence of extreme weather conditions described here is one of those associated with the change in

Table 3.2 Monthly indices of precipitation and temperature for England for 1310–29

Precipitation

Year	Dec.	Jan.	Feb.	Mar.	Apr.	May	Jun.	Jul.	Aug.	Sep.	Oct.	Nov.
1310						-2	-2	-2			-2	-2
1311						-2	-2	-2				
1312				-2		-2		-2			-2	-2
1313	(2)	(2)	(2)					-2		2	2	2
1314							2	2	2	2	2	3
1315	2	2					3	3	3			3
1316			3			3		2		3	3	
1317	2	2				2		2		2	2	2
1318						2	-3					
1319							-3		(2)	3	3	
1320					2					2	2	
1321				2	2		-2	-2				
1322					(2)				(2)	(2)	(2)	
1323					(2)		-2				-2	-2
1324									(2)	(2)	(2)	
1325												
1326	-2	-2	-2				-3	-3	-3			
1327		-2					-2	-2				
1328												
1329	(2)	(2)	(2)				-3	-3				

Temperature

Year	Dec.	Jan.	Feb.	Mar.	Apr.	May	Jun.	Jul.	Aug.	Sep.	Oct.	Nov.
1310		-3									-2	-2
1311			(-2)									
1312												
1313		-2	-2		-2							
1314		-2	-2									
1315												
1316												
1317		-2	-2									
1318						-2						
1319												
1320												
1321												
1322			-3	-3	-3							
1323												
1324												-2
1325												
1326		-2	-2	-2	-2							
1327												-2
1328		-2	-2	-2								
1329												-2

Source: Farmer and Wigley 1983
Note: The figures in parentheses are less firmly established

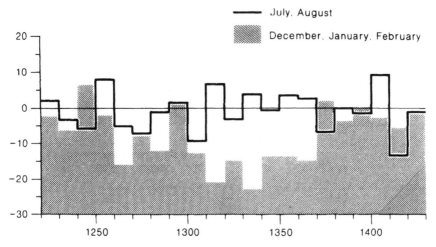

Figure 3.19 Decadal indices for 1220–1450, showing summer wetness in June and July, and winter severity in December, January and February. Index values were obtained by summing all the relevant monthly totals for each decade.
(From Farmer & Wigley 1983)

general climatic conditions in the early part of the Little Ice Age. It certainly had many and diverse effects on English agriculture and society. Naturally these were also influenced by other factors operating at the same time, including war with Scotland, the growth of towns, and the introduction of diseased livestock from France.

The Little Ice Age centuries: sixteenth- and seventeenth-century Crete

The Mediterranean basin, as well as lands farther north, was affected by climatic stresses during the Little Ice Age. It is possible to trace both the incidence of climatic anomalies and some of their results from papers and letters written by the Venetian administrators of the island (Grove and Conterio 1992). Between 1548 and 1648 Crete experienced many severe winters and prolonged winter and spring droughts (Figure 3.20). The latter were particularly serious from the agricultural point of view and caused an extraordinary predominance of inadequate or failed harvests (Table 3.3).

The island of Crete was of great strategic and no doubt psychological importance to the Venetians. By the sixteenth and seventeenth centuries Venetian power was waning in the east as that of the Ottoman Turks rose, and the essential provisioning of the island presented very great difficulties. Its predominant importance was clearly recognised by administrators, such as those who wrote to the Doge in 1588:

Since the main task that Your Serenity gave to those who represent him

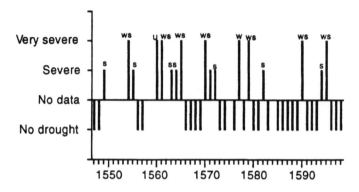

(a)

(b)

Figure 3.20 The incidence of droughts in Crete 1547/8 and 1644/5.
(From Grove and Conterio 1995) (a) 1547/8–1598/9 (b) 1599/1600–1644/5
w = winter, s = spring, u = unspecified

in his towns is that they should make sure that there is sufficient grain for those who live there and that when his territories do not yield the quantity that is needed, they should make sure that some is brought from other places. So, knowing that this territory does not produce enough of what is needed for the whole year, as soon as we arrived we hurried to find out about the current situation.

(Ludovic Memmo, Duke of Candai, and Antonio Miani,
Captain General. 14.10.1581. ASV.PTM. Filza 746)

In fact yields were good in 1581, increasing by as much as twenty- or thirty-fold for barley. But such good years were very infrequent (Table 3.3). At this time the incidence of extremes of climate was much greater than during the present century, especially as regards droughts in both winter and spring and

Table 3.3 Climate and crops in Crete, 1548–1648

Year	Winter	Spring	Summer	Crops, sirica, *plague and famine*
	– normal	– normal	– normal	b barley + good
	x severe	+ favourable	+ good	c cereals – normal
	X very severe	x unfavourable	x unfavourable	w wheat – poor
	D drought	D drought		v grapes x failure
				o olives
1547–8	–	x	–	c– v+
1548–9	–	+		
1549–50		D		c– famine
1550–1				famine
1551–2				famine
1552–3				
1553–4				
1554–5	D Nov, Dec, Jan, Feb	D Mar, rain in May	x water shortage	c– w– famine
1555–6	x	D April	x	c– famine and emigration
1556–7	–	+		c–
1557–8	–			
1558–9				o–
1559–60				
1560–61	Drought at unspecified time			c–
1561–2	D	D		c–
1562–3				Rethimnon c– Sitia and Candia cx
1563–4		D in May		c– famine
1564–5	–	D in March	x water shortage	c– famine
1565–6	D hot until late Nov	D	x water shortage	c–
1566–7	x wet			
1567–8	X wet			c–
1568–9	–	x	x	cx
1569–70	x wet	x		bx c–
1570–1	D	D April showers	–	b– w– v+
1571–2	Drought at unspecified time		Turkish invasion	
1572–3		D south wind in April	x N and NW winds and rough seas in August	c– w– v– in most places but Lassithi and Messerea w–
1573–4	–	+	x	b+ but w– on account of *sirica*
1574–5	–	+	+	c+ w+ v– o– but cx in Canea
1575–6	x cold and wet		x rain June, July, Aug	c– w– v–
1576–7	x	+	x	w– *sirica*
1577–8	D			c–

Year	Winter	Spring	Summer	Crops, sirica, plague and famine
	– normal x severe X very severe D drought	– normal + favourable x unfavourable D drought	– normal + good x unfavourable	b barley + good c cereals – normal w wheat – poor v grapes x failure o olives
1578–9	–	+	x	c– w– *sirica* but Candia w–
1579–80	D but Spinalonga –	D in Mar, April south winds in April	x grazing sparse	w– o+ but wx in Rethimnon
1580–1	–	+		b– w– o–
1581–2		+	– but wet June	c+ o–
1582–3	–	D south winds		b– w–
1583–4	–	+		b+ w– especially in Messerea and Lassithi
1585–6	x 15 days of July temperatures in Jan			b+ w– ox
1586–7	x snow and rain			b– w– but Omalos w+ Canea w–
1587–8	–	+	–	b+ w– o [?]
1588–9	X snow	=	–	w–
1589–90	D Nov, Dec, Jan, Feb	D water shortage	x deluge in June	c– v– famine, *sirica* in Lassithi
1590–1	X cold and wet, deluge in Rethimnon and in Feb	+	NW winds in July	b+ w– o+
1591–2	X cold and snow	+	strong NW winds in August	b+ w– v– plague
1592–3	–	+	–	o– w+ v+ but plague reduces manpower for harvest
1593–4	– but x in Spinalonga	=	x south wind deluge in August	b– w– v– plague
1594–5	X cold and snow seed corn eaten	x D in April and May		w– v+ plague
1595–6	D Nov, Dec, Jan no forage, animals die	x D in April and May	x very hot streams and cisterns dry	b– w– v– famine
1596–7	–	+	=	b– w+ o–
1597–8	– wet January	x		c– o+
1598–9	–	–		w–

Table 3.3 continued

Year	Winter	Spring	Summer	Crops, sirica, *plague and famine*
	– normal	– normal	– normal	b barley + good
	x severe	+ favourable	+ good	c cereals – normal
	X very severe	x unfavourable	x unfavourable	w wheat – poor
	D drought	D drought		v grapes x failure
				o olives
1599–1600	–		very hot June	
1600–1	D Sept until January			o–
1601–2	X extreme cold	D rain in May	–	b– v–
1602–3	X cold and snow	x	x	c– *sirica*
1603–4	–	D		c– w– o– but Lassithi w+
1604–5	X excessive rain	+	+	w+
1605–6	–	+ south winds in May	+	b– v– but Lassithi w+ v+
1606–7	–	D		c–
1607–8	–	+	–	b+ c–
1608–9	X excessive rain and snow	+		
1609–10	–	+	–	v+ o–
1610–11	–	+	very hot	
1611–12	–	+	+	c+
1612–13	– south winds in Oct	x		c–
1613–14	D in Dec, Jan, Feb, Mar	D April	x	c– w–
1614–15	–	+ but seed lacking	x rainstorms	b+ c– owing to lack of seed
1615–16	–	+		
1616–17				
1617–18	X deluge Dec 28	+		b– c–
1618–19				
1619–20			June hot	o–
1620–1		x	x	b– w– o– famine
1621–2	–	May storm		famine
1622–3	– but stormy			w+ o+
1623–4		+	+	b+ w+ o–
1624–5	–	x dry	x south winds in July springs dry	c– but c– in Lassithi v– o–
1625–6	D Jan, Feb, March	D April forage scarce	x	c– v– o– Lassithi w– Some peasants leave the countryside
1626–7	–	–		

Year	Winter	Spring	Summer	Crops, sirica, plague and famine
	– normal	– normal	– normal	b barley + good
	x severe	+ favourable	+ good	c cereals – normal
	X very severe	x unfavourable	x unfavourable	w wheat – poor
	D drought	D drought		v grapes x failure
				o olives
1627–8			+	c+ but Canea c–
1628–9	–			b– w–
1629–30	–	– south wind in April	south winds in June	b– w+
1630–1	D	D rain in May		w– Lassithi w– v– o–
1631–2	–	+	x July thunderstorms damage v & o	b+ w– v– o–
1632–3	–	+		
1633–4	–	– April gales		
1634–5	x	stormy		
1635–6				w– c– o+
1636–7		+		
1637–8	x stormy	+	c– especially in Sithia →	[?]
1638–9	x snow	D April		
1639–40				
1640–1		x		w–
1641–2				w–
1642–3	–	D south winds in May	x south winds in June	c– w– v– famine in Canea. Seed corn eaten
1643–4				c– v+ famine
1644–5	X widespread deluge			
1645–6				
1646–7		almost all Crete occupied by Turks		
1647–8			NW gales in Aug	

Source: Grove and Conterio, 1995
Note: The condition the Venetians knew as sirica was probably some kind of plant disease that attacked cereals, particularly wheat but never barley. It caused the grains to blacken and rot.

severe winters. Yields were generally inadequate; sometimes so inadequate as to cause famine, as in 1595 and 1621–3 for instance. In all the years of spring drought both the barley and wheat crops failed or were very poor except in 1631, when some of the barley was saved by timely rain, which broke the drought temporarily.

Bad weather could harm the crops at almost any stage of their growth. A long, severe winter could delay sowing, as in 1595 when in March it was reported that the farmers had been forced to eat their seed. Drought in the autumn or winter months could have the same effect as in 1596. Over-much rain in winter could be as serious as too little. Three months' continuous rain in 1605 prevented autumn cultivation as well as causing great damage to the walls of the fortresses on the island. Excessive rain was even more unwelcome when it was accompanied by severe cold, as in 1602 when 'because of the continuous rain and extreme cold a large number of animals have died on Crete'. That winter the cold was also held responsible for many deaths among the crews of the galleys on guard duty.

Heavy rain in summer, such as the storm of 21 July 1632 which ruined many vines and olive trees, could also be serious. A sequence of hazards could produce particularly unfortunate effects, as in 1590, when drought, which 'reduced the cereals to straw', was followed by heavy rain on 1 June which 'destroyed and burnt the greater part of the vines' and also uprooted many trees. But of all the climatic hazards suffered by the Venetians in Crete, drought associated with the prevalence of south winds was the most frequent. The situation described by Zuanne Giacomo in 1614, when the continual south winds 'have not only caused a drought of four months' duration, but have also been very detrimental to the sown fields', was unfortunately not uncommon. Even if conditions in winter and spring were good, south winds could blight hopes of a good harvest in a very short time, as they did in 1625:

> The harvest this year ... has turned out to be very different from what was expected and from the hope that was first conceived of it. The south winds, which have reigned for only two or three days, have diminished it so much that it will be barren rather than mediocre.
> (Girolomo Trevisan. 30.7.1625. ASV.PTM. Filza 783)

Shortfall in production involved serious financial burdens and inherent political instability. When the crops were insufficient Venice either had to send more supplies than usual, or provide money for stores to be bought from foreign sources. The underlying problem of the decision to hold an island so far away was frankly stated in what turned out to be an accurate forecast:

> Habitual shortage of grain in the realm was something difficult to cope with in peacetime without grain from enemy countries, but in time of war it is impossible.... The Turk knows that in case of war he can in the long term impoverish the island to the point where it will fall into his hands.
> (Alvise Grimani. 22.11.1585. ASV.PTM. Filza 752)

Contact between Crete and Venice depended on the links in the long route down the Adriatic and from island to island (Figure 3.21). Shipwrecks were all too frequent, causing loss of ships and cargoes and long delays in

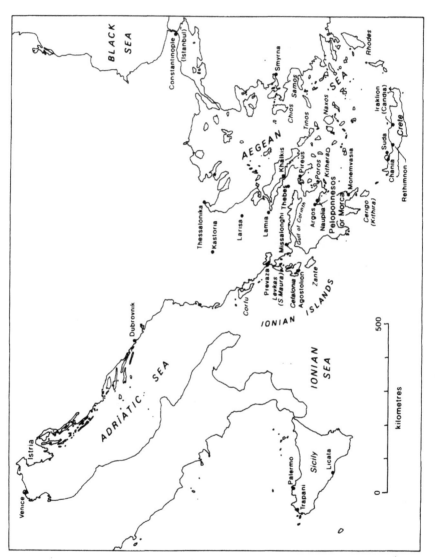

Figure 3.21 The position of Crete in the eastern Mediterranean

communications. Governor-General Nicolo Donado spent six months of his thirty-nine in post getting to and fro and suffered two shipwrecks on his return voyage. Shipwrecks were not infrequently multiple. Most costly of all was the terrible storm which 'caused the loss of most of the light fleet ... all the surviving ships amount to the number of eight ... all the rest are believed to have been lost' (Luca Francesco Barbaro. 21.3.1658. ASV.PTM. Filza 808). By this time much of Crete was already occupied by the Turks. Ten years earlier the people of Candia had waited in vain for 'compassionate succour' which could come only by sea. It was the Cretan peasants, on whose labour home-produced food supplies depended, who suffered most severely during the many years of shortage. In the worst of these they had to depend on famine foods such as carobs and wild plants. Mortality was high in the worst of the famines, especially among children. But even the troops suffered from cold and the high prices of food during shortages, as they had to buy their own, with inadequate salaries.

In order to obtain foreign supplies the Cretan administrators had to have sufficient finance. Repeated letters to Venice begging for supplies to be sent were interspersed with requests for money. Purchase of grain was either from Turkish vessels, which frequently came to the Cretan ports in the years of relative peace, or from the Aegean islands or the Greek mainland. Again and again ships were sent out to search for supplies, which were often not available, either because the Turks were banning sales or because of pirates. But just as often the trouble was that no surpluses were to be had because of crop failure over a much wider area, including large parts of the eastern Mediterranean. The Turks themselves suffered from famines during this period.

There were factors other than the weather which limited Cretan food production. Shortage of labour could occur during the most serious of the plague epidemics. Then there was the wish of the landlords to concentrate on growing vines. Cretan wine was well established in several European markets and brought in much better returns than cereal growing, especially as the Venetians made the mistake of controlling cereal prices. Several attempts were made to diminish the area under vines by uprooting or preventing expansion but these were not very successful. It is arguable that a different pricing policy would have had better results. Another reason was the attitude of the landlords to the peasants, whose difficulties some of the administrators came to appreciate:

> They cultivate only what little piece of land they think enough to provide for their basic needs. However, if the peasant could be assured of the continued protection of the Republic so as to give him the slightest hope of increasing the yield for himself and his sons rather than the landowners, then he would be industrious, hardworking and willing ... and try his best to improve his affairs, with real benefit for

the public good and the improvement of agriculture. In some territories
... the owners, in lending money to the tenants for agricultural needs,
have recourse to usury, with interest of 150% per year.
(Alvise Gustinian 1591. ASV.Collegio Relazione, B.79cc.15v–16r.)

There were many factors leading to the loss of Crete by the Venetians
among which climate should not be overlooked. Climatic extremes were
common throughout the final century of the Venetian occupation, and
presented repeated difficulties. Food supplies were already short in Italy in
the 1580s. In 1590 Venice had to buy grain from Danzig for home use
(Braudel 1972, p. 600). The long series of poor Cretan harvests, the hard
conditions, particularly those suffered by the troops during the long final
siege of Candia, and the loss of ships at sea were surely a factor in the eventual
loss of the island to the Turks.

CONCLUSIONS

Climate is only one of the variables which influence human societies. Its
influence is most likely to be strong in areas marginal for agriculture, but
even there the nature of the interplay between climatic change and society
depends crucially on the adaptability of the group involved. The contrasting
cases of the Norse in Greenland and the Anasazi in Colorado illustrate this
very clearly, as does also that of southeast Scotland. One of the main century-
scale episodes of the Holocene, the Medieval Warm Period, influenced all
three of them. It was also almost certainly one of the factors which promoted
expansion of population in England, a country enjoying a temperate climate
much more conducive to agricultural success.

Century-scale fluctuations consist of sequences of decadal and smaller-
scale variations. The Anasazi reacted positively to a series of such fluctuations during the warm period. The Norse, failing to adjust to the combined
impact of deteriorating climate and related biological factors at the beginning
of the Little Ice Age, together with social and economic factors involved in
the cessation of contact with the outside world, were unable to survive,
unlike their Inuit neighbours. Farmers in Scotland who had been able to
expand arable to higher altitudes during the benign years had to desert their
fields when their crops failed too often.

Thirteenth- and fourteenth-century English society was far more complex
socially and economically than those of the areas so far discussed. Here the
impact of adverse weather conditions in the early Little Ice Age was bound
to have much more intricate consequences, ranging from death by starvation
at one end of the scale to the acquisition of riches at the other. It could not
have a common influence upon different social and economic groups, upon
the grain and the wool trade or even upon all regions of the country.
Evaluation of the impact therefore cannot be simple but should not be

ignored. This case illustrates some of the complexities of dealing with incomplete data relating to a complex situation. Like that of southeast Scotland, it draws attention to the possibility that a series of environmental fluctuations may constitute a gradual decline in the resource base.

Crete in the sixteenth and seventeenth centuries played a role in an extensive network of long-distance trading, as well as in a long-running political power tussle between very distinct regional powers having different social and religious values. It cannot be argued that the series of decadal weather fluctuations experienced towards the end of Venetian rule had a conclusive influence on political or economic life. But that it had rather more than a slight influence would seem very probable.

The impact of climatic fluctuations on technologically advanced countries has been much cushioned since about 1850, as a result of the industrial revolution, with cheap transport allowing rapid movement of bulk goods and modern agricultural techniques including plant breeding. However, many less economically advanced regions still suffer from the effects of climatic fluctuations, especially if they are accompanied by war, poverty, and inadequate administration. Droughts in the Sahel in recent decades provide an obvious example.

The impact of climatic changes on human society depends upon their scale on the one hand and the characteristics and sensitivity of the society concerned on the other. Here we might notice that the widespread climatically induced grain shortages in the Mediterranean in the 1590s may well have assisted the rise of the Netherlands to entrepreneurial power. The decadal changes of the Little Ice Age centuries have been noted and their effects are worthy of study on a wide as well as a local scale (e.g. Post 1980). The impact of conditions in a single year, 1816 for instance, can be considerable (Harington 1992). Direct effects are less noticeable in modern advanced societies. But even here the impact of, for instance, a very warm or cold winter will have economic consequences which may be very important for some sectors of society (Manley 1957). International trading patterns may well again be affected by climatic oscillations, as they were in 1972.

There is currently a great deal of concern about the probability of future climatic oscillations and speculation about their possible effects (IPCC 1990, 1992). Public attention has been drawn to the possibility of changes on a scale that would produce effects that could not be ignored even by rich, technologically advanced societies. In considering these matters the nature of past climate has to be taken into account, for the natural factors which caused fluctuations in the past have not ceased to operate and would have appreciable moderating or magnifying effects on any anthropogenically induced changes. Climatic as well as social and economic forces must affect human history, and will continue to do so.

NOTES

1 This pattern has been associated with some rather unfruitful disagreement about the timing of the Little Ice Age in Europe. It can be argued that it is more useful to think about the history of climate from the point of view of dynamic processes rather than static conditions (Fischer 1981).

2 Pfister's study of Little Ice Age climate in Switzerland led to the formation of the CLIMHIST data bank. This is now the focus of the work of a large number of European investigators employing the same methodology, which is already making it possible to construct historical weather maps, revealing the sequence of past synoptic patterns.

3 Retrodiction of the effect of climatic fluctuations on past crop yields uses analogous data on the effect of weather on crop yields at present, allowing for differences in these relationships with developments in crop strains over time.

4 Titow (1961–2) calculated that the arable land in the hands of the peasant of Taunton in 1248 was 3.3 acres per person and that in 1311 it was at best 2.5 acres; that is, there was by then acute land shortage. He reckoned that the increase in the area held by the peasant must have come from the demesne lands. The area of the demesne under seed decreased between 1248 and 1311 by some 640 acres, which with the three-field system represented an overall decrease of 960 acres.

5 An exception is J. D. Post's comment (1980, p. 722) that while the population of Europe grew only slowly during the seventeenth and eighteenth centuries (except in the Netherlands and England), 'this was striking when contrasted with the rapid population growth during the twelfth and thirteenth centuries, when the European climate was apparently milder and less variable'.

6 This article is based on a very large number of unverified sources, but there seems little doubt that the general findings may be safely accepted, although details may be incorrect.

7 No good sources survive to provide close estimates of urban death rates in England, but in Flanders the burial register of the city of Ypres shows that 2,794 persons, estimated as one-tenth of the total population, died between May and October 1316 (Kershaw 1976).

8 An infectious livestock disease.

9 An increase in wind storms, causing flooding of low-lying coasts, was associated with the general deterioration of climate (Lamb 1977, p. 452). A study of loss of coastal land, based on original sources, was made by Bailey (1993). Unfortunately he did not attempt to map his findings.

10 The cattle murrain may well not have been associated with climatic conditions, though it might be argued that beasts which had had inadequate feed in the preceding years may have been more vulnerable to disease.

11 Out of nearly fifty deserted villages in Oxfordshire and Northamptonshire for which adequate population figures of tax-paying tenants are available, all declined seriously between 1311 and 1319 and the numbers were down by a third compared with those for 1280-1300. The decline by 1327 averaged 67 per cent in those places later wiped off the map by the Black Death (Lamb 1977, p. 455).

12 Even the Winchester data present some difficulties of missing data and terminology, concerned, for instance, with the definition of customary acres.

13 A few years later combined meetings of historians and climatologists led to the publication of seminal studies on both sides of the Atlantic (Rotberg and Rabb 1981; Wigley et al. 1981) which began to fill the gaps in understanding between disciplines.

14 It is noteworthy that Postan accepted and used the outdated climatic chronologies of Britton (1937) and Brooks (1949), which are now known to include

important errors. Unfortunately, early paleoclimatologists frequently failed to appreciate the necessity for rigorous examination of the historical accuracy of sources, while historians, examining their chronologies and finding them of no historical significance, failed to understand the cumulative and progressive nature of scientific research.

Archival references

ASV = Archivio di Stato Venezia

PMT = Provveditori da Terra e da Mar

REFERENCES AND FURTHER READING

Abel, W. 1980. *Agricultural Fluctuations in Europe from the Thirteenth Century to the Twentieth Centuries.* Methuen.

Bailey, M. 1993. Per impetum maris, natural disaster and economic decline in eastern England, 1275–1350. In Campbell, B. M. S. (ed.) *Before the Black Death: Studies in the 'Crisis' of the Early Fourteenth Century.* Manchester: Manchester University Press, 184–208.

Beresford, G. 1981. Climatic change and its effect uppon the settlement and desertion of medieval villages in Britain. In Delano Smith, C., and Parry, M. (eds) *Consequences of Climatic Change.* Department of Geography, University of Nottingham, 30–9.

Blikra, L., and Nemec, W. 1993. Postglacial avalanche activity in western Norway: depositional facies sequences, chronostratigraphy and paleoclimatic implications. In Frenzel, B. (ed.), Matthews, J.A., and Glaser, B. (co-eds) *Solifluction and Climatic Variation in the Holocene.* Stuttgart: Gustav Fischer Verlag, 143–62.

Bond, G., Broecker, W., Johsen, S., Mauns, J., Labeyreie, L., Jouzel, J., and Bonani, G. 1993. Correlations between climatic records from north Atlantic sediments and Greenland ice. *Nature* 364: 218–20; 365: 143–7.

Bradley, R. S., and Jones, P. D. 1992. *Climate since AD 1500.* London and New York: Routledge.

Bradley, R. S., and Jones, P. D. 1993. 'Little Ice Age' summer temperature variations: their nature and relevance to recent global warming trends. *Holocene* 3: 367–76.

Braudel, F. 1972. *The Mediterranean and the Mediterranean World in the reign of Philip II.* 2 vols. London: Collins (translation by Sian Reynolds of 2nd revised edition).

Briffa, K. R., and Jones, P. D. 1993. Global surface air temperature variations during the twentieth century: Part 2, implication for large-scale high-frequency paleoclimatic studies. *Holocene* 3: 82–93.

Britton, C. E. 1937. *A Meteorological Chronology to AD 1450.* Meteorological Office Geophysical Memoir 70. London: HMSO.

Brooks, C. E. P. 1949. *Climate through the Ages: A Study of the Climatic Factors and their Variations,* revised edn. London: Benn.

Burns, B. T. 1983. Simulated Anasazi storage behaviour using crop yields reconstructed from tree-rings: AD 652–1968. Unpublished Ph.D. dissertation, Department of Anthropology, University of Arizona, Tucson.

Campbell, B. M. S. (ed.) 1993. *Before the Black Death: Studies in the 'Crisis' of the Early Fourteenth Century.* Manchester: Manchester University Press.

Cordell, L. S. 1984. *Prehistory of the Southwest.* New York: Academic Press.

Dansgaard, W., Johnsen, S., Reach, N., Gundersdrup, N., Clausen, H. B., and Hammer, C. U. 1975. Climatic changes, Norsemen and modern man. *Nature* 255: 24–8.

Delibrias, G., Le Roy Ladurie, M., and Le Roy Ladurie, E. 1975. La forêt fossile de Grindelwald: nouvelles datations. *Inter-Sciences*, 137–47.

Euler, R. C., Gumerman, G. J., Karlstrom, T. N. B., Dean, J. S. and Hevly, R. H. 1979. The Colorado Plateaus: cultural dynamics and paleoenvironment. *Science* 205: 1089–101.

Farmer, G., and Wigley, T. M. L. (eds). 1983. *The Reconstruction of European Climate on Decadal and Longer Time-Scales*. Final Report. Commission of the European Communities. Contract No. CL-029-81-UK(H).

Fischer, D. J. 1981. Climate and history: priorities for research. In Rotberg, R. I., and Rabb, T. K. (eds) *Climate and History: Studies in Interdisciplinary History*. Princeton NJ: Princeton University Press.

Gad, F. 1970. *The History of Greenland. I. Earliest Times to 1700* (translated from Danish by E. Dupoint). London: Hurst.

Grootes, P. M., Stuiver, M., White, J. W. C., Johnsen, S., and Jouzel, J. 1993. Comparison of oxygen isotope records from the GISP2 and GRIP Greenland ice cores. *Nature* 366: 5524.

Grove, J. M. 1972. The incidence of landslides, avalanches and floods in western Norway during the Little Ice Age. *Arctic and Alpine Research* 4: 131–8.

Grove, J. M. 1988. *The Little Ice Age*. London: Methuen.

Grove, J. M., and Battagel, A. 1983. Tax records from western Norway as an index of Little Ice Age environmental and economic deterioration. *Climatic Change* 5: 265–82.

Grove, J.M., and Conterio, A. 1992. Little Ice Age climate in the eastern Mediterranean. In: Mikami, T. (ed.) *Proceedings of the International Symposium on Little Ice Age Climate*. Tokyo: Tokyo Metropolitan University, 221–6.

Grove, J. M., and Conterio, A. 1994. Climate in the eastern and central Mediterranean 1675–1715. *Paleoklimaforschung: Paleoclimatic Research*, Special Issue 13.

Grove, J. M., and Conterio, A. 1995. The climate of Crete in the sixteenth and seventeenth centuries. *Climatic Change* 30: 223–47.

Grove, J. M. and Switsur, R. 1994. Glacial geological evidence for the Medieval Warm People. *Climactic Change* 26(2/3), March.

Harington, C. R. (ed.). 1992. *The Year without a Summer: World Climate in 1816*. Ottawa: Canadian Museum of Nature.

Harvey, B. F. 1993. Introduction: the crisis of the early fourteenth century. In Campbell, B. M. S. (ed.) *Before the Black Death: Studies in the 'Crisis' of the Early Fourteenth Century*. Manchester: Manchester University Press, 1–24.

Harvey, P. D. A. 1976. The English inflation of 1180–1220. In Hilton, R. H. (ed.) *Peasants, Knights and Heretics: Studies in Medieval English Social History*. Cambridge: Cambridge University Press, 57–84.

Holzhauser, H. 1984. Rekonstruktion von Gletschwankungen mit Hilfe fossiler Hölzer. *Geographica Helvetica* 39: 3–15.

IPCC [International Panel on Climate Change]. 1990. *Climatic Change: The IPCC Scientific Assessment*. World Meteorological Organization/United Nations Environmental Programme.

IPCC [International Panel on Climate Change]. 1992. *Climate Change 1992: The Supplementary Report to The IPCC Scientific Assessment*. Cambridge: Cambridge University Press.

Jones, P. D. 1990. The climate of the past 1000 years. *Endeavour*, n.s. 14: 129–36.

Kerr, R. A. 1993. The whole world had a case of the Ice Age shivers. *Science* 262: 1072–3.

Kershaw, I. 1976. The agrarian crisis in England 1315–1322. In Hilton, R. H. (ed.) *Peasants, Knights and Heretics: Studies in Medieval English Social History.* Cambridge: Cambridge University Press, 85–132.

Lamb, H. H. 1977. *Climate: Present, Past and Future*, vol. 2. London: Methuen.

Lamb, H. H. 1984. Climate in the last thousand years: natural climatic fluctuations and change. In Flöhn, H., and Fantechi, R. (eds) *The Climate of Europe: Past, Present and Future.* Dordrecht: Reidel, 25–64.

Lorius, C., and Oeschger, H. 1994. Paleo-perspectives reducing uncertainties in global change. *Ambio* 23: 30–6.

Lucas, H. S. 1962. The great European famine. In Carus Wilson, E. M. (ed.) *Essays in Economic History*, vol. 2. London: Edward Arnold, 49–72.

Lütschg, O. 1926. *Über Niederschlag und Abiluss im Hochgebirge.* Sonderstellung des Maltmorkgebietes. Schweizerische Wasserwirtschattverband, Verband schriff No. 14. Veröffentlichung der Schweizerischen meteorologischen Zentralanstalt in Zürich. Secrétariat de l'Association Hydraulique Suisse.

McGovern, T. H. 1981. The economics of extinction in Norse Greenland. In Wigley, T. M. L., Ingram, M. J., and Farmer, G. (eds) *Climate and History: Studies in Past Climates and their Impact on Man.* Cambridge: Cambridge University Press, 404–33.

Manley, G. 1957. Climatic fluctuations and fuel requirements. *Scottish Geographical Magazine* 73: 19–28.

Mate, M. 1993. The agrarian economy of south-east England before the Black Death: depressed or buoyant? In Campbell, B. M. S. (ed.) *Before the Black Death: Studies in the 'Crisis' of the Early Fourteenth Century.* Manchester: Manchester University Press, 79–109.

Messerli, B., Messerli, P., Pfister, C., and Zumbühl, H. J. 1978. Fluctuations of climate and glaciers in the Bernese Oberland, Switzerland and their geoecological significance, 1600 to 1975. *Arctic and Alpine Research* 10: 247–60.

Nichols, H. 1974. Arctic North American palaecology: the recent history of vegetation and climate deduced from pollen analysis. In Ives, J. D. and Barry, R. J. (eds) *Arctic and Alpine Environments.* London: Methuen, 637–67.

Nichols, H. 1975. Palynological and paleoclimatic study of the late Quaternary displacement of the forest–tundra ecotone in Keewatin and Mackenzie, N.W.T., Canada. Occasional paper 15, Institute of Arctic and Alpine Research, University of Colorado.

Parry, M. L. 1975. Secular climatic change and marginal land. *Transactions of the Institute of British Geographers* 64: 1–13.

Parry, M. L. 1976. The significance of the variability of summer weather in upland Britain. *Weather* 31: 212–17.

Parry, M. L. 1978. *Climatic Change, Agriculture and Settlement.* Folkstone: Dawson.

Parry, M. L. 1981. Climatic change and the agricultural frontier. In Wigley, T. M. L., Ingram, M. J., and Farmer, G. (eds) *Climate and History: Studies in Past Climate and their Impact on Man.* Cambridge: Cambridge University Press, 319–36.

Parry, M. L., and Carter, T. R. 1985. The effect of climatic variation on agricultural risk. *Climatic Change* 7: 95–110.

Petersen, K. L. 1988. Climate and the Dolores River Anasazi. Anthropological Papers 113. University of Utah, Salt Lake City.

Pfister, C. 1978. Climate and economy in eighteenth century Switzerland. *Journal of Interdisciplinary History* 10: 719–23.

Pfister, C. 1981. An analysis of the Little Ice Age climate in Switzerland and its consequences for agricultural production. In Wigley, T. M. L., Ingram, M. J., and Farmer, G. (eds) *Climate and History: Studies on Past Climates and their Impact on Man.* Cambridge: Cambridge University Press, 214–48.

Pfister, C. 1988. Fluctuations climatiques et prix céréaliers en Europe du XVIe au XXe siècle. *Annales ESC*, Jan.–Feb., 25–53.

Pfister, C. 1992. Monthly temperature and precipitation in central Europe 1525–1979: quantifying documentary evidence on weather and its effects. In Bradley, R. S., and Jones, P. D. (eds) *Climate since AD 1500*. London and New York: Routledge, 118–42.

Post, J. D. 1977. *The Last Great Subsistence Crisis in the Western World.* Baltimore: Johns Hopkins University Press.

Post, J. D. 1980. The impact of climate on political, social and economic change: a comment. *Journal of Interdisciplinary History* 10: 719–23.

Postan, M. M. 1975. *The Medieval Economy and Society*, 2nd edn. Harmondsworth: Penguin.

Postan, M. M., and Titow, J. Z. 1959. Heriots and prices on Winchester manors. *Economic History Review*, 2nd series 11: 392.

Reynaud, L. 1980. Can the linear mass balance model be extended to the whole Alps? In Muller, F. (ed.) *World Glacier Inventory: Proceedings of the Reideralp Workshop, 1978.* International Association of Hydrological Sciences, Publication 126, 273–84.

Rotberg, R. P., and Rabb, T. K. 1981. *Climate and History: Studies in Interdisciplinary History.* Princeton, NJ: Princeton University Press.

Röthlisberger, F. 1974. Études des variations climatiques d'après les histoires des cols glaciaires: le Col de Herens (Valais Suisse). *Bollettino del Comitato Glaciologico Italiano*, 2nd series 22: 9–34.

Röthlisberger, F. 1986. *10,000 Jahre Gletschwankungen der Erde.* Aarau: Verlag Saulander.

Schlanger, S. H. 1988. Patterns of population movement and long-term population growth in southwestern Colorado. *American Antiquity* 53: 773–93.

Shackleton, N. J., and Opdyke, N. D. 1973. Oxygen isotope and paleomagnetic statigraphy of equatorial Pacific core V28-238. *Quaternary Research* 3: 39–55.

Slatter, E. D. 1979. Drought and demographic change in the prehistoric Southwest United States: a preliminary quantitative assessment. Unpublished Ph.D. dissertation, Department of Anthropology, University of California Los Angeles.

Taylor, K. C., Hammer, C. U., Alley, R. B., Clausen, H. B., Dahl-Jensen, D., Gow, A. J., Gundestrup, N. S., Klipfstuhl, J., Moore, J. C., and Waddington, E. D. 1993. Electrical conductivity measurements from GISP2 and GRIP Greenland ice cores. *Nature* 366: 549–52.

Titow, J. Z. 1959–60. Evidence of weather in the account rolls of the bishopric of Winchester 1209–1350. *Economic History Review*, n.s. 12: 360–407.

Titow, J. Z. 1961–2. Some evidence of thirteenth century population increase. *Economic History Review*, n.s. 14: 218–23.

Titow, J. Z. 1972. *Winchester Yields: A Study in Medieval Agricultural Productivity.* Cambridge: Cambridge University Press.

Venetz, I. 1833. *Mémoire sur les variations de la température dans les Alpes de la Suisse.* Written in 1821. Zurich: Orelli Fussli.

Vibe, C. 1978. Cyclic fluctuations in tide related to season as key to some important short and long term fluctuations in climate and ecology of the North Atlantic and Arctic regions. In Frydenddahl, K. (ed.) *Proceedings of the Nordic Symposium on Climatic Changes and Related Problems.* Danish Meteorological Institute Climatological Papers 4, Copenhagen.

Wigley, T. M. L., and Kelly, P. M. 1990. Holocene climatic changes: ^{14}C wiggles and variations in solar irradiance. *Philosophical Transactions of the Royal Society of London Series A* 330: 547–60.

Wigley, T. M. L., Ingram, M. J., and Farmer, G. (eds). 1981. *Climate and History: Studies in Past Climates and their Impact on Man.* Cambridge: Cambridge University Press.

Zumbühl, H. J. 1980. *Die Schwankungen der Grindelwaldgletscher in den historischen Bild- und Schriftquellen des 12. bis 19. Jahrhunderts.* Zurich: Denkschriften der Schweizerischen Naturforschenden-Gesellschaft.

IDENTIFYING THE TIME-SCALES OF ENVIRONMENTAL CHANGE

The instrumental record

Nicholas J. Clifford and John McClatchey

Editors' note The comment is often made that 'environmental change' is frequently used synonymously with 'climatic change'. In this chapter Nicholas Clifford and John McClatchey attempt to see if we can identify changes of a shorter duration than Grove considered. They do so with respect to one variable, temperature in England, because we have comparatively good records for this variable over the past three hundred years. But even in examining one variable it rapidly becomes clear that there are many indices that may be relevant - including for example the frequency of temperature extremes rather than averages. Their major conclusions throw doubt on our ability either to understand or to predict even short-term change, and these caveats should be borne in mind when in later chapters there are passing and sometimes rather token references to 'scientific uncertainty'.

INTRODUCTION

Time-scales of environmental change and the instrumental record

Long homogeneous series of instrumental climatic data are crucially important in examining past climate changes. Not only do these data provide a direct record of climate, and possible climatic shift over the inter-decadal scales most relevant for economic and social activity (see Grove, Chapter 3), but they are also needed to calibrate proxy data such as those gleaned from documentary records (Bradley and Jones 1992). More recently, attention has also focused on the use of instrumental series for assessing the plausibility of alternative scenarios predicted by climatic models. With the increased recognition of the possibility of global warming and/or synoptic shifts in climatic phenomena since the later nineteenth century, there has been added impetus given to the study of instrumental records for establishing not just the reality of climatic change, but also its nature. This requires that steady-state or various forcing scenarios need to be viewed alongside 'random' variation within the present climatic regime (Probert-Jones 1984).

The longest and most robust instrumental series, and hence one of the most important, is that of the monthly mean Central England temperature (CET) developed by Manley (1953). This chapter examines alternative means of visualising and identifying the time-scales of change in environmental systems based upon this series. The greater part of the discussion is based upon the quantitative analysis of the time series data using standard econometric analysis in which the time series is examined for trend, seasonal oscillation, and random variation. Comparisons are drawn between this approach and other statistical procedures, and some preliminary thoughts are given concerning more novel analysis which aims to elucidate the nature of the underlying system dynamics. While a great deal of emphasis has been placed on warming *trends*, the chapter demonstrates that it may be more appropriate to look for a variety of inhomogeneities in climatic records including elements of pseudo-cyclic and 'noise' behaviour. Individual time-scales of variation should always be identified and compared with respect to longer- and shorter-term frames of reference, and the sources of variability considered carefully in the light of the graphical and statistical procedures being used. Trends may, for example, be more apparent than real where series values are smoothed as part of any analysis. One goal in the examination of instrumental records is to model the variations which they exhibit, and this may be approached statistically or via physically based numerical simulation. Some success with statistical modelling is reported here, and, in view of the possible nature of environmental time series as chaotic (or at least, extremely complex), this may prove the most fruitful avenue of research into the future.

The Central England temperature series

The Central England temperature series represents the outcome of a series of publications concerning monthly mean temperatures for the period 1659–1973, which were united by Manley (1974). The construction of the series presented considerable problems (Manley 1953), as most observations before 1841 were the work of enthusiastic amateurs. The observational records were therefore widely scattered, of diverse length, and of varying accuracy. Construction of the series was also complicated by the range of relief across the country which affects the observations at individual sites. It was these features of the records which led Manley (1953, p. 245) to believe that 'the best method of standardising any of scattered earlier English records would be to try to bring them all to a comparable standard, namely a "Central England mean"'. At that time this mean was represented by records from Radcliffe (Oxford) plus 'Lancashire' divided by two.

In 1953, Manley described the Central England mean as being the monthly twenty-four-hour mean which would derive from a site of intermediate character between 30 and 45 m (100–150 ft) in Shropshire, south Cheshire or Worcestershire. In a later paper (Manley 1974), it was suggested the series would most closely match the results of observations from stations of intermediate character in open rural surroundings in the lowlands of Staffordshire, Shropshire and north Warwickshire, at 30–90 m (100–200 ft) above sea-level. Manley (1974) explained that the term 'intermediate character' implied that CET was not representative of stations located in frost hollows, on wide stretches of sandy soil, or on exceptionally windswept ridges.

Manley (1974) also provided a clear indication of the reliability of the CET data at different periods over the 315 years from 1659 to 1973. He clearly stated that the estimates for the first six decades to 1720 were less reliable, and to emphasise this the data for those decades were presented in italics. In addition, the data for the months from 1659 to 1670 were quoted to the nearest whole degree Celsius and to the nearest 0.5 °C from 1671 to 1698 and again from 1707 to (October) 1722. In all other months, the data are given to the nearest 0.1 °C. The later paper (Manley 1974) provided improved estimates of the CET series over the period 1699–1706 based on a reconsideration of observations from London and improvements of the estimates of CET for 1707–1722. For the latter period, Manley had to make use of observations from Utrecht in the Netherlands (Labrijn 1945) plus some information from 'wind and weather' diaries, as no English temperature records were available.

Since 1974, the UK Meteorological Office has updated the monthly CET series and constructed a daily series back to 1772 (Jones 1987; Parker et al. 1992). This daily series was adjusted to ensure that monthly mean temperatures from its daily values matched Manley's CET monthly values, and is now used for updating Manley's monthly series (Parker et al. 1992). Although Manley took care to avoid urban sites, the construction of the daily series has indicated that some of the stations used since 1973 are liable to progressive warming due to urbanisation. Corrections to monthly values have been made from 1980 onwards, therefore. From 1983, all monthly values have been corrected by –0.1 °C, and from 1992 by –0.2 °C in June and July (Parker et al. 1992).

CHARACTERISING RECORDS, AND EXAMINATION OF TRENDS

The most frequent question with respect to climatic variation concerns the issue of linear trend: is the world warming, if so by how much, and over what time-scale? However, logically, as Dyer (1976) points out, the idea of linear trend is meaningless, unless it is assumed that the trend continues for ever.

More important questions (and ones which have some greater physical rationale) concern the *homogeneity* of environmental time series: the degree of variance over time displayed by a variety of statistical measures, which incorporate pseudo-cyclic changes, changes in extremes, and random fluctuations. Often, these questions are also of more direct relevance to the impact on human socio-economic systems (see below). Although climatic (and especially environmental) change is necessarily complicated, this complexity should not be seen as overwhelming, since what is needed is a general framework for intelligent assessment of possible human modifications compared to intrinsic variability (Dickinson 1986). Given this context, while there is doubt as to the appropriate extrapolation of precise details from any given instrumental record, its use is more easily justified with respect to the likelihood that it will disclose the kinds of past change, and offer insights into the most appropriate ways of predicting and representing future changes.

Establishing a frame of reference: the longer term

An appropriate long-term frame of reference for instrumental climatic records (10^2 years) can be established with respect to geological time periods ($>10^4$ years), and time-scales below this from several centuries into the present. Broadly, our knowledge derived from a range of proxy data (ice and sediment cores, tree-rings, and, for historical periods, documentary sources) reveals that over geological time periods, large amounts of variability occur in pseudo-cycles of approximately 20,000, 40,000, and 100,000 years. These can be related to earth orbital parameters (for reviews of evidence and mechanisms see Schneider and Londer 1984) and effectively set a background context of 'intrinsic variability' against which contemporary fears of anthropogenetically modified and rapid time-scales of changes must be set.

Importantly, geological variation seems to have been relatively large given the modest variation in solar radiation flux (although it has remained within bounds acceptable for some form of life over very long periods), and this is indicative of internal (possibly regulatory) feedback between land, ocean, and biosphere systems. Whatever the particular time-scale of interest, therefore, this comparison must always be made against several other possible scales and kinds of variation because of this complexity (McElroy 1986). Figure 4.1 illustrates temperature change over the past 18,000 years, which is probably short enough a period to exclude major changes in external solar forcing. Here, two periods of considerable significance for human activity – the Medieval Warm Epoch and the Little Ice Age – occurred immediately after one another, but both were confined to mean temperature shifts of only *c.* 1.5 °C, compared with shifts at least three times greater since the last glacial maximum. Over these time-scales, therefore, the intrinsic capacity for change in the world's climate appears to be a dominant consideration.

The shorter term: rates and magnitudes of change and statistical homogeneity

Figure 4.1 also reveals the importance of considering the rate and duration of climatic change as well as its magnitude. Thus, long-period trends show changes of *c.* 0.02 °C per century, but much higher rates have occurred naturally for shorter periods. The Little Ice Age, for example, involved a cooling of about 1 °C over two hundred years (see Grove, Chapter 3). For comparison, the Central England temperature series is shown in Figure 4.2, with a centre-line of the overall series mean (9.2 °C) marked for convenience. Simple linear regression fits for the entire series yield a coefficient of 0.002 (i.e. 0.2 °C per century), while a regression for the period 1880–1992 gives a coefficient of 0.006, or 0.6 °C per century. There is an apparent upward trend, particularly during the twentieth century, with enhanced warming from the late nineteenth century onwards. However, this is still less than has occurred in the historical past, and very much less than in the recent geological past without any possibility of human influence (compare with Figure 4.1). Given Dyer's (1976) comments on the inadequacy of trends to conceptualise climatic variation, what may be more useful, therefore, is a more complex view of the relationship between trend, the distribution of the statistics comprising this trend, and patterns within the series such as cyclicity or 'runs' of similar values.

Several attempts have been made to assess the homogeneity of the CET.

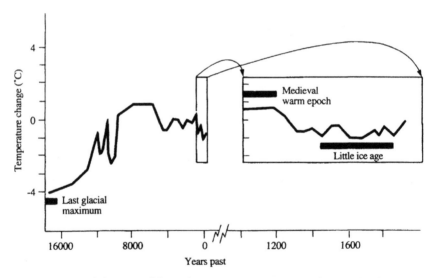

Figure 4.1 Variation in mid-latitude mean temperature over the past 18,000 years. Inset shows the two most recent phases of 'major' climatic variation relevant to human activity, but note the nesting of the time-scales (magnitude and duration) of natural changes. (After Parry 1986)

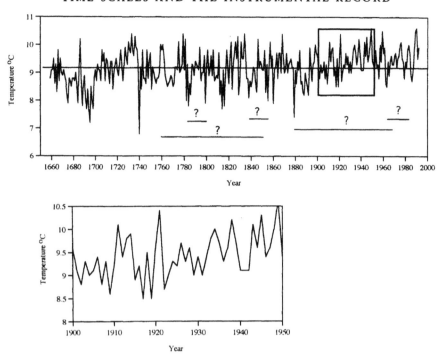

Figure 4.2 The Central England temperature series, 1659–1992. Series is plotted together with its long-term mean of 9.2°C. Inset illustrates 'pulse-like' variation and an apparently enhanced upward-trend to the series from the late nineteenth century onwards. Bars on the diagrams represent approximate 90-year and 20–25-year 'cycles' following analysis by Dyer (1976)

These have been based upon 'distribution' analyses, which seek to examine the fit of the entire series to some single, assumed frequency distribution model, and/or analyses which examine the statistical properties of successive arbitrary periods within the series. Homogeneity is shown if the series corresponds to the assumed frequency distribution, and/or where successive periods show greater statistical overlap than their internal differences. Probert-Jones (1984) found negative skewness (–0.37) in annual means which he attributed to an early cold phase, but the entire distribution was well fitted by a Pearson Type IV frequency function, suggesting homogeneity. This was reinforced by a combination of autocorrelation and regression analysis of annual means and their standard errors, which demonstrated residual values from a regression model to be Gaussian (i.e. normally distributed). An analysis of means over 10- to 50-year periods again supplied no evidence of statistical change in the occurrence of values over the *entire* period, but it is, perhaps, noteworthy that all the 10-year means from 1901 have been above the long-term average (although not above the 95 per cent confidence interval). Results from a similar exercise are shown in Figure 4.3 and Table 4.1, where boxplots of annual means are calculated for standard World Meteorological Organization (WMO) 30-year periods. Boxplots are a particularly useful tool since they summarise graphically both the values within, and the structure of, frequency distributions: the box is essentially the middle half of a distribution with the upper and lower quartiles comprising a box which encloses the median. 'Whiskers' reach out to illustrate the range of data, and more extreme 'outlier' values are represented by stars (Velleman and Hoaglin 1981). The case for considering changing *distributions* of statistics when assessing the importance of environmental variability for human activity is presented by Parry (1986), who emphasises that it is sensitivity to extreme or 'threshold' events which is likely to be a more relevant definition of climatic change. Figure 4.4 illustrates that a change in the frequency of extremes may occur with or without a change in the mean of a distribution, depending upon the presence or absence of a shift in 'level' as part of a climatic trend or a more abrupt transition. A changing frequency of extremes may be considered as a change in risk of impact, and also a change in resource opportunities, as some opportunities are closed off and new ones emerge. With respect to the CET, the results are particularly striking: mean and median temperatures vary between 8.8 and 9.6 °C for the periods listed in Figure 4.3 and Table 4.1, with standard deviations in the range 0.5–0.8 °C. Maximum temperatures for the same periods, however, show a very narrow range from 10.1 to 10.6 °C, whereas minimum

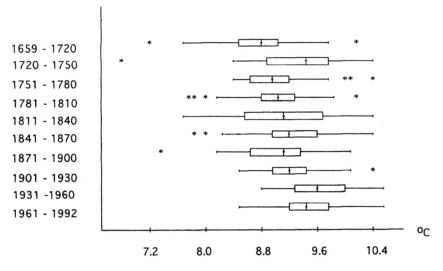

Figure 4.3 Boxplots of the statistical properties of the Central England temperature series for the period 1659–1720 and for successive WMO standard 30-year periods. For explanation of the boxplots see text

Table 4.1 Characteristics of the Central England temperature series – entire period and standard WMO 30-year-periods

Period	Mean	Median	Standard deviation	Maximum	Minimum
1659–1992	9.2	9.2	0.6	10.6	6.8
1659–1720	8.8	8.8	0.6	10.2	7.2
1721–1750	9.4	9.5	0.7	10.4	6.8
1751–1780	9.1	9.0	0.5	10.4	8.4
1781–1810	9.0	9.1	0.6	10.2	7.8
1811–1840	9.1	9.2	0.8	10.4	7.7
1841–1870	9.2	9.2	0.6	10.4	7.9
1871–1900	9.1	9.2	0.6	10.1	7.4
1901–1930	9.3	9.3	0.5	10.4	8.5
1931–1960	9.6	9.6	0.5	10.6	8.8
1961–1992	9.5	9.5	0.5	10.6	8.5

Note: All figures are in °C to one decimal place, but see text for caution regarding differential accuracy in several time periods

temperatures exhibit a much larger range of variation, from 6.8 to 8.8°C. Reference back to Figure 4.2 in conjunction with the boxplots reveals that it is the past sixty years which shows the clearest signs of sensitivity in these respects, with a marked reduction in the extremes of minimum temperatures.

Although this analysis has been performed for the annual temperature series, it is important to note that the annual statistics may themselves reflect more complex changes in the seasonality of monthly series. In a detailed investigation of secular trend, Dyer (1976) demonstrated that blocks of months behave differently: summer and part of autumn were stationary in their statistical properties, whereas later autumn, and winter and spring are undergoing upward variation. Craddock (1965) and Kendall and Stuart (1966) used principal components analysis to suggest secular movements, in addition to seasonal and annual effects, as being responsible for this apparent trend. Seasonal changes, therefore, are manifested in the annual statistics as a slight upward, and possibly non-linear, component. Recognition of seasonal and other 'sublevel' variability in mean statistics is vital if instrumental records are to be used for providing realistic scenarios for risk and other assessments (such as calibration or scenarios for climatic models) in addition to their use for identifying the existence and time-scales of change. In particular, while equilibrium climatic models are run until an 'end-point' scenario is reached, dynamic models should reveal changes in temperature through time which could be compared to seasonal and subseasonal variation from the instrumental record. Nevertheless, a variety of problems remain with such attempts, and some of these are discussed below.

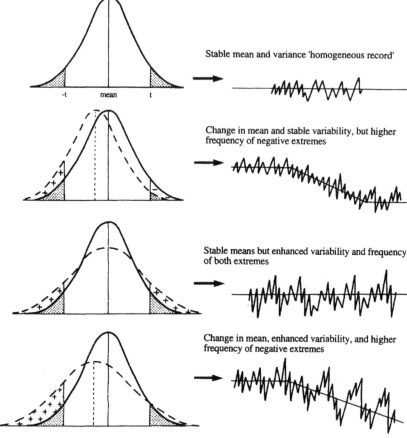

Figure 4.4 Characterising climatic shifts with respect to changes in a variety of statistical measures. On each diagram, *t* represents a notional 'threshold' value of the climatic parameter which has significance for human activity, such as agricultural yields. Changes in the level (mean) of the distribution create more or less 'threshold' events, as do changes in the shape of the distribution. (Based on Parry 1986, after Fukui 1979)

THE INSTRUMENTAL RECORD: PROBLEMS AND PROSPECTS

Time and space scales and the representation of change

Although the CET is widely recognised as one of the most robust instrumental series, it still provides an example of the difficulties in making inferences concerning the time-scales of environmental change from the record of a single environmental variable. Ideally, changes in the series should be related to changes in the physical environmental system of the place or region where they originate, but these dual requirements of physical and spatial representativeness are not necessarily consistent, and both are difficult to assess.

With respect to possible physical reasons for fluctuations in the CET, autocorrelation and spectral analysis of monthly means show possible periodicities of 94, 20–25 and *c.* 2 years (Dyer 1976). Reference back to Figure 4.2 provides some support for this, where approximate variations of 90- and 25-year duration are suggested. Very short-term spectral peaks have been identified at quasi-biennial, 11- and 22-year variations in a variety of other climatic series (Rasmusson *et al.* 1981), and some of these are thought to be associated with sunspot activity. However, a causal linkage is unclear (Pittock 1978). An expanded section of the CET from 1900 to 1950 (inset to Figure 4.2) is suggestive of elements of 2- to 5-year variation, but again, both the physical interpretation and the significance of these are clouded by recognition of another common phenomenon in climatic series: 'red noise' elements. These arise from superposition of 'pulses' which are 'turned on' at random points in time and are thought to relate to the internal dynamics of the atmosphere and its coupling to other elements of the climatic system (see Dickinson 1986). Short-term weather features are known to cause much variability in monthly averages and even year-to-year variation at a given location (Leith 1973), so that an apparent time-dependent structure may exist even without a systematic cause.

Another problem in correctly identifying the appropriate representation of change may arise when the skewness or kurtosis of a distribution, rather than change in its level (mean), is involved. Such changes in the frequency distribution can pose especially difficult problems for identification of longer-term behaviour, because of the sensitivity of some smoothing procedures. In the case of instrumental records a degree of smoothing or filtering has been seen as desirable in order to suppress high-frequency fluctuations and hence emphasise longer-term variation (Mitchell 1966), and smoothing has been extensively employed as an essential element in the comparison between differing series. Thompson (1995), for example, uses low-pass filtering in an attempt to clarify the relationships between six European instrumental series in an examination of changing continentality over the historical period. However, as with series means, linear low-pass (smoothing) filters suffer from an inherent sensitivity to skewness: if not all individual observations are presented, potentially few 'anomalies' can create the appearance of a broader change having taken place (Barring and Mattsson 1992).

In order to illustrate the advantages and disadvantages of filtering, a simple 9-year moving average was used to smooth the CET, before fitting a high-order polynomial regression to the smoothed data (Figure 4.5). On first impression, there appears to be a change towards higher-frequency variation in the latter part of the series post-*c.* 1900, together with a marked decrease in the amplitude of this fluctuation. In particular, the early (pre-*c.* 1750) record shows more marked temperature extremes. In addition, a low-period cyclicity may be evidenced from the polynomial, although, given the narrow temperature range which it encompasses, its significance is rather dubious (see also below on accuracy of the unfiltered data). However, reference back to the boxplots in Figure 4.3 shows that this simple two-part distinction is also somewhat misleading: there is a reduction in outlier observations post-*c.* 1800, and some very low anomalies also occur in that earlier period, but median temperature values are not clearly consistent with either an over-all warming trend or a simple twofold classification. For comparison, the lower bars in Figure 4.5 refer to qualitiative 'periods' or phases identified in a similar temperature series for Uppsala by Barring and Mattsson (1992) which are supposedly three 'epochs': 1722–*c.* 1830, characterised by high-amplitude, shorter-term fluctuations with a fairly constant longer-term average; 1830–1930, with much lower amplitude variation and a longer-term cool phase; and 1930–*c.* 1989, with a restabilisation to a higher-temperature longer-term mean and renewed higher-amplitude oscillation. Clearly, there is no correspondence between the series in these respects. While this might at first lead to adverse conclusions regarding the ability to compare series and produce regional-scale climatic 'test' or calibration scenarios for global circulation models, Barring and Mattsson caution that their 'epochs' are largely artefacts of the low-pass filtering procedures they employed! Once alternative filtering methods (based upon median filters and a novel 'running histogram' comparison) were used, their second cooler epoch was almost non-existent, and was so close in absolute terms to the median estimates in

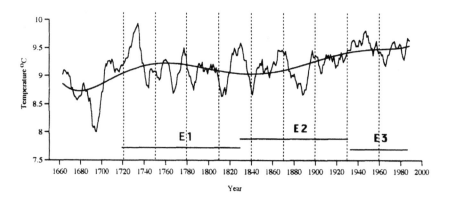

Figure 4.5 The Central England temperature series for 1659–1992 illustrating the effects of a 9-year running mean together with a high-order polynomial regression. Smoothing is designed to emphasise underlying variation, but is sensitive to changes in the frequency distribution of data within subperiods (see Figure 4.3). For comparison, E1–E3 refer to three apparent 'epochs' identified in a similar temperature series for Uppsala, Sweden (Barring and Mattsson 1992). For further discussion, see text

earlier periods that serious questions concerning spurious comparison need to be considered because of doubts concerning data reliability. Similar caution should be applied here to the CET. Although use of the CET series has been extensive (for example, Schonwiese 1978, 1980; Murray 1992) and it is still commonly used as a standard for comparisons in Europe (for example, Pfister 1992), there is often no reference to Manley's (1974) caution about the reliability of the early data or of the rounding of those data to the nearest 0.5 or 1°C in certain periods. Probert-Jones (1984), for example, ascribed a standard error of 0.2°C to annual means and 0.1°C to decadal means before 1723, yet still quotes other statistics to two decimal places! The lack of explicit consideration of within-series variation in accuracy may also complicate conclusions arising from comparison of highly smoothed data in Thompson's (1995) recent study.

Some alternative methods of assessing the homogeneity of time series are shown in Figure 4.6. In Figure 4.6a, a time series of annual temperature differences is shown. This clearly highlights non-homogeneity of series variance, since there appears to be almost a wave- or pulse-like pattern as well as instances of individual much larger extreme values. Figures 4.6b–4.6f attempt to clarify the nature of this behaviour in the form of directed scatter plots for arbitrary time periods. The directed scatter plots show successive annual deviations from the long-term mean plotted against one another. Plots in the upper right quadrant, for example, are warmer years than the average which have been preceded by a warmer year. Plots in the upper left quadrant show cooler than average years which have been preceded by a warmer one. The overall scatter is a measure of extremes in the time period, and the centroid of the plot is a measure of the *tendency* in a period compared to the long-term mean: a centroid displaced from the origin into the upper right quadrant is thus indicative of an overall warmer 'phase'. The figures add new insights into the debates outlined above concerning trend and anomalies: there is no clear separation between an earlier cooler and a later warmer period, although pre-1900 plots do show greater extremes in minimum values. What is more striking, however, is the changing overall shape of the plots from periods of more even scatter into all quadrants (1741–1900 and especially 1900–60) to plots where there is a marked asymmetry (1659–1740 and 1960–92). Asymmetry in the plots shows a strong preference for 'runs' of similar temperature behaviour – warmer preceded by warmer and cooler followed by cooler – and the shape and degree of scatter is a prelude to judging the possible dynamical behaviour of the underlying climatic system, as discussed below in the final section.

Figures 4.2–4.6 of the CET provide ample illustrations of the need to identify trends and time-scales of behaviour from several statistical measures, and to exercise particular caution when using mean values. To some extent, this is unfortunate given the care taken in past research to provide series which are most robust in the mean values, and the emphasis on low-pass

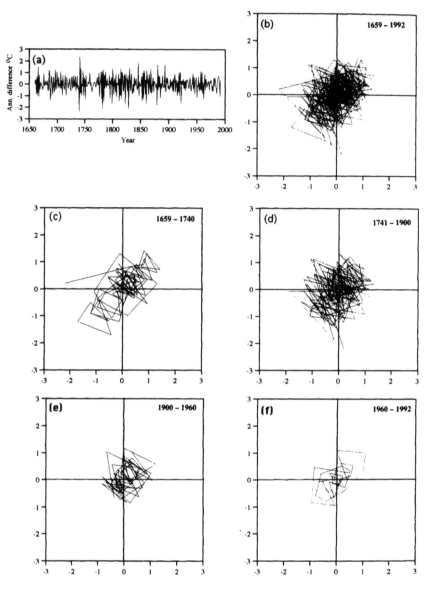

Figure 4.6 An alternative way of representing change and homogeneity in environmental time-series. Figure 4.6a shows annual differences in the Central England temperature series plotted through time. Figures 4.6b–4.6f are directed scatter plots, which show deviations about the overall series mean plotted against the preceding year's deviation. Both the shape and positioning of the plots within the four quadrants are important clues to series homogeneity. For further explanation, see text

filtering to identify longer-term trends. Inevitably, therefore, there will be many instances where conflicting results are obtained depending upon the statistical measure adopted. In the face of these difficulties, both alternative conceptualisations of change are required, and, equally important, alternative means of differentiating between the time-scales of change. In particular, the insights added by the directed scatter plots strongly suggest that the autocorrelation properties of the series may change over time, which in turn may indicate some sort of self-adjustment mechanism in the dynamics of that part of the climatic system which series such as the CET represent. In the sections which follow, two such alternatives are introduced. The first is an attempt to model series behaviour as a stochastic process based upon the autocorrelation properties of the series; the second is an attempt to gain an insight into the nature of the variability, and hence provide theoretical limits to series predictability and its associated time-scales of change.

Complex behaviour: modelling trend, cyclicity, and random variation

Figure 4.7 illustrates the application of a statistical modelling technique based upon ARIMA (autoregressive integrated moving-average) models. Box and Jenkins (1976) provide the classic reference on this subject, but more accessible treatments are given from a general perspective by Chatfield (1984) and from a geographical/environmental perspective by Richards (1979). Harvey (1993) also provides an excellent account of the relationship between analysis in the time and frequency domains, and the specifics of the statistical procedures used to generate these results can be found in Vandaele (1983) and the *MINITAB Reference Manual* (Minitab Inc. 1991).

As a standard econometric technique, ARIMA modelling is used as a complement to the 'trend regression' approach. ARIMA models are similar in structure to regression models, but are applied where the only variable is some function of itself evolving through time plus a random 'disturbance' term. Although it is a stochastic approach (a one-dimensional data series is assumed to evolve through time as the outcome of a statistical process), it is intrinsically more dynamic than regression modelling and is therefore able to capture more elements of the behaviour of potentially inhomogeneous series. In addition, applications in the environmental sciences have exploited the flexibility of the stochastic model to represent a physically determined (but complex and possibly poorly specified) system operating in the presence of one or many noise sources, which is again attractive when considering the nature of both climatic systems and the instrumental record.

The time series used in ARIMA modelling is assumed to be approximately stationary and normally distributed with constant mean and variance, although in practice these assumptions can often be relaxed in view of possible de-trending and transformation bias. Moving-average (MA) models

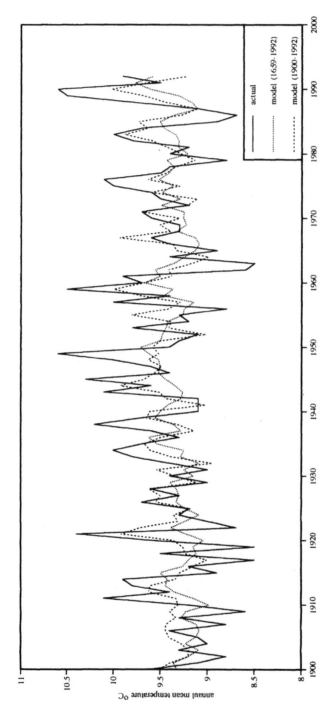

Figure 4.7 The results of ARIMA modelling of the Central England temperature series. Results are shown for the application of two models: a model for the entire series 1659–1992, and a second model for the series 1900–92. Comparison over the post-1900 period shows good agreement for the position of peaks and troughs for both models, but the shorter-term model is clearly better. Both models underrepresent the actual variability in the series, although in all but two years this variability is within the 95 per cent model confidence limits. For further discussion, see text

incorporate past random fluctuations ('shocks') to represent the time series and are represented generally by

$$Z_t = a_t - f_1 a_{t-1} - f_2 a_{t-2} - \ldots f_q a_{t-q}$$

where f_1 and f_2 are the MA coefficients and a is the random 'shock' term.

Autoregressive (AR) models estimate values of the dependent variable as a regression function of previous values, generally given as

$$Z_t = f_1 Z_{t-1} + f_2 Z_{t-2} + \ldots f_q Z_{t-q} + a_t$$

Importantly, it is possible to mix models since mixed models can be written in terms of pure models of infinite order (Vandaele 1983). Mixed models are often desired when examining complex series, since these provide more parsimonious model fits (fewer model parameters have to be estimated) than pure AR or MA models alone. Practically, therefore, a mixed structure can provide additional flexibility where an output series may result from more than one interacting process (Salas *et al.* 1980). The models used here are multiplicative seasonal ones with the general form ARIMA:

$$(p, d, q)\,(P, D, Q)\,S$$

where lower-case symbols represent the non-seasonal component (p = order of the AR part; d is the order of series differencing; and q is the order of the MA part) and upper-case ones the seasonal component with a seasonality of S (= 12 for an annual trend in monthly data). This is only one way of coping with seasonality, of course, but it avoids the prior removal of trends and cylicity which would then have to be added back to later forecasts of the non-seasonal components. One model, therefore, represents many aspects of series behaviour.

Two ARIMA models were fitted using annual mean temperature data:

- for the entire series 1659–1992 the form is: (1 0 1) (0 0 1) 15
- for the series 1900–92 the form is: (2 0 2) (1 0 1) 10

and the results of each model fit are compared to actual data from the CET in Figure 4.7. In both cases, models were identified according to three criteria based upon the modelling framework developed by Box and Jenkins (1976): first, the appropriate model is identified according to the form of the autocorrelation and partial autocorrelation functions; second, the coefficients of the model are estimated based upon minimisation of residual variance; and third, the Q statistic, or the so-called Portmanteau test, is used, which tests the null hypothesis that the residual autocorrelations are associated with a random series. Note that in all cases, ordinary least-squares procedures are used to provide best-fit models, and that model components are expressed in multiplicative form, which allows much simpler evaluation of stationarity conditions. No differencing terms were employed since the data were expressed as residuals about the long-term and post-1900 series mean values.

There are several points of note when considering Figure 4.7. First, with respect to the entire CET (1659–1992), the model is less complex (i.e. low order) than for the period 1900–92 in both seasonal and non-seasonal parts. This would partly be expected, since a much longer series is being fitted, and there is a higher chance of what, effectively, are random elements of behaviour. Second, both the simplicity of the full CET model and especially the longer seasonal component associated with it support the conclusions from Figures 4.2 and 4.3, which indicate lower-amplitude and higher-frequency behaviour in the latter part of the series, post-c. 1910. Third, both models underrepresent the variance of the actual series, and fits from the model of the entire series are much poorer. Finally, the full CET model also demonstrates a possible shift in phase in many instances, although, considering the length and complexity of the series, the results are surprisingly good, and the phase shift does not appear to be systematic. In general, the correspondence of many peaks and troughs in the actual and fitted series, particularly that from 1900, is very encouraging, and suggests that the approach is worth pursuing further, possibly incorporating the monthly temperature values.

Inevitably, the question arises as to the possibility of forecasting using this approach. Although there may be possibilities in this respect, it is important to remember that stochastic modelling by definition cannot simulate the physical processes behind the series, merely mimic their statistical character-istics. At best, the limits to predictability will, therefore, be heuristic. What is really required is to determine the underlying nature of observed variation. Does the data series represent randomness superimposed upon determinism, or is there chaotic behaviour, where complicated behaviour is actually simple in origin (random in appearance but with a small number of degrees of freedom)? As Wilcox et al. (1991) note, there has been a flurry of recent activity searching for chaos in environmental time series. The most popular technique is to search quantitatively for chaos in a time series signal based upon the correlation and geometric properties of the time series represented in a similar manner to Figure 4.6. Applications of this analysis to short-term weather phenomena are given by Fraedrich (1987), Essex et al. (1987), Tsonis and Elsner (1988), and Rodriguez-Iturbe et al. (1989); and to longer-term climatic series by Nicolis and Nicolis (1984) and Grassberger (1986). However, more recent applications of these techniques highlight the absence of definite criteria for series pre-processing and interpretation of results, particularly where the environmental time series is highly autocorrelated, and a major limitation is the length of data required (for discussion, see Theiler 1986; Wilcox et al. 1991). Some of these issues are treated further by Clifford and McClatchey (in prep.), where a comparison between results from the annual CET and monthly CET is presented. At best, however, the results are tentative and conjectural, and are particularly restricted with respect to the annual series. Given these restrictions, no further discussion is attempted here, since the essential point of debate – complexity versus possible chaos – is still unresolved.

CONCLUSIONS

Analysis of the CET data illustrates many of the potentials and limitations in attempts to identify the time-scales of environmental change from instrumental climatic series. Taken at face value, instrumental series present perhaps the greatest opportunities to assess climatic variability in many of its guises over time-scales relevant to human activity. However, the very nature of climatic and environmental change can be an obstacle to the representation of change. Despite growing attempts to extend the use of instrumental series to disclose the reality of enhanced climatic shifts, and to compare between series (possibly as a prelude to providing climatic scenarios to evaluate climatic models), there are some very basic questions which remain to be answered regarding the accuracy of data, which frequently alter within series in a non-systematic way.

In addition, it is important to account for changes which are non-linear in origin and which are not solely (if at all) associated with changes in the level (mean) of a distribution. There are strong theoretical grounds for paying more attention to the frequency of extremes when assessing environmental change over time-scales relevant to socio-economic activities, but these changes in extremes may easily be misrepresented as shifts in means and climatic 'phases', especially where linear low-pass smoothing is applied. By contrast, a preliminary attempt to model the CET using stochastic time series models is encouraging, offering a statistical means of encompassing several aspects of climatic variability within a single model.

Conceptually, however, perhaps the most important question still to be answered from an examination of instrumental records is concerned with the fundamental (physical) nature of observed variation. If it does turn out that such series can quantitatively be shown to display low-dimensional chaotic behaviour, then the future use of the instrumental record is likely to be severely restricted. Sensitivity to initial conditions inevitably leads to a poor capacity for predicting future behaviour. An appropriate, if disquieting, analogy in this instance may be Wigley's (1986) idea of climatic change as an unrolling carpet: a new carpet is rolling out under our feet as we travel through time, but the carpet of past climate is actually rolling up behind us as we go! Judged against this metaphor, then, scenarios of contrasting time-scales of environmental change may well be less relevant than ones of constant evolution.

ACKNOWLEDGEMENTS

The authors would like to thank Professor E. C. Zeeman for his encouraging comments in respect of their efforts to analyse the nature of variability in environmental time series. N. J. C. would also like to acknowledge the influence of Dick and Jean Grove, who introduced him to the subject of climatic change and to the importance of historical and instrumental environmental records.

REFERENCES

Barring, L., and Mattsson, J. O. 1992. Influence of anomalous years on filtered time series of the annual temperature from Uppsala, Sweden. *Geografiska Annaler* 74A: 275–82.

Box, G. E. P., and Jenkins, G. M. 1976. *Time Series Analysis, Forecasting and Control*, revised edn. San Francisco: Holden-Day.

Bradley, R. S., and Jones, P. D. 1992. Climate since AD 1500: introduction. In Bradley, R. S., and Jones P. D. (eds) *Climate since AD 1500*. London: Routledge, 1–16.

Chatfield, C. 1984. *The Analysis of Time Series: An Introduction*. New York: Chapman and Hall.

Clifford, N. J., and McClatchey, J. (in prep.) Stochastic and dynamic aspects of environmental time series. To be submitted to *Physical Geography*.

Craddock, S. M. 1965. The analysis of meteorological time series for use in forecasting. *Statistician* 15(2): 167–90.

Dickinson, R. E. 1986. Impact of human activities on climate – a framework. In Clark, W. C., and Munn, R. E. (eds) *Sustainable Development of the Biosphere*. Cambridge: Cambridge University Press, 252–89.

Dyer, T. G. J. 1976. An analysis of Manley's central England temperature data: I. *Quarterly Journal of the Royal Meteorological Society* 102: 871–88.

Essex, C. T., Lookman, T., and Nerenberg, M. A. H. 1987. The climate attractor over short time-scales. *Nature* 326: 64–6.

Fraedrich, K. 1987. Estimating weather and climate predictability on attractors. *Journal of Atmospheric Sciences* 44(4): 722–8.

Fukui, H. 1979. Climatic variability and agriculture in tropical moist regions. *Proceedings of the World Climate Conference*, WMO No. 537. Geneva: World Meteorological Organization, 311–18.

Grassberger, P. 1986. Do climatic attractors exist? *Nature* 323: 609–12.

Harvey, A. C. 1993. *Time Series Models*, 2nd edn. London: Harvester Wheatsheaf.

Jones, D. E. 1987. Daily Central England temperature: recently constructed series. *Weather* 42: 130–3.

Kendall, and Stuart, 1966. *The Advanced Theory of Statistics*. London: Charles Griffin.

Labrijn, A. 1945. The climate of the Netherlands during the last two and a half centuries. *Mededeelingen en Verhandelingen* 49, KNMI No. 102.

Leith, C. E. 1973. The standard error of time-average estimates of climatic means. *Journal of Applied Meteorology* 12(6): 1066–9.

McElroy, M. B. 1986. Change in the natural environment of the Earth: the historical record. In Clark, W. C., and Munn, R. E. (eds) *Sustainable Development of the Biosphere*. Cambridge: Cambridge University Press, 199–211.

Manley, G. 1953. The mean temperature of central England, 1698–1952. *Quarterly Journal of the Royal Meteorological Society* 79: 242–61.

Manley, G. 1974. Central England temperatures: monthly means 1659 to 1973. *Quarterly Journal of the Royal Meteorological Society* 100: 389–405.

Minitab Inc. 1991. *MINITAB Reference Manual*, Minitab Inc.

Mitchell, J. M. 1966. Climatic change. WMO Technical Note No. 79.

Murray, R. 1992. Some notable features of Manley's Central England mean temperatures with special reference to very warm years. *Weather* 47: 98–103.

Nicolis, C., and Nicolis, G. 1984. Is there a climatic attractor? *Nature* 311: 529–32.

Parker, D. E., Legg, T. P., and Folland, C. K. 1992. A new daily Central England temperature series, 1772–1991. *International Journal of Climatology* 12: 317–42.

Parry, M. L. 1986. Some implications of climatic change for human development. In Clark, W. C., and Munn, R. E. (eds) *Sustainable Development of the Biosphere*. Cambridge: Cambridge University Press, 378–407.

Pfister, C. 1992. Monthly temperature and precipitation in central Europe 1525–1979: quantifying documentary evidence on weather and its effects. In Bradley, R. S. and Jones, P. D. (eds) *Climate since AD 1500*. London: Routledge, 118–42.

Pittock, A. B. 1978. A critical look at long-term sun–weather relationships. *Reviews of Geophysics and Space Physics* 16: 400–16.

Probert-Jones, J. R. 1984. On the homogeneity of the annual temperature of central England since 1659. *Journal of Climatology* 4: 241–53.

Rasmusson, E. M., Arkin, P. A., Chen, W.-Y., and Jalickee, J. B. 1981. Biennial variations in surface temperature over the United States as revealed by singular decomposition. *Monthly Weather Review* 109: 587–98.

Richards, K. S. 1979. *Stochastic Processes in One-Dimensional Series*. CATMOG 23, Geo Absracts, Norwich.

Rodriguez-Iturbe, I., Febres de Power, B., Sharifi, M. B. and Georgakakos, K. P. 1989. Chaos in rainfall. *Water Resources Research* 25(7): 1667–75.

Salas, J. D., Delleur, J. W., Yevjevich, V., and Lane, W. L. 1980. *Applied Modelling of Hydrologic Time Series*. Littleton, CO: Water Resources Publications.

Schneider, S. H., and Londer, R. 1984. *The Co-evolution of Climate and Life*. San Francisco: Sierra Club Books.

Schonwiese, C. D. 1978. Central England temperature and sunspot variability 1660–1975. *Archiv für Meteorologie, Geophysik und Bioklimatologie Ser. B*, 26: 1–16.

Schonwiese, C. D. 1980. Statistical comparison of central England annual and monthly air temperature variability, 1660–1977. *Meteorological Magazine* 109: 101–13.

Theiler, J. 1986. Spurious dimension from correlation algorithms applied to limited time-series data. *Physics Reviews A* 34(3): 2427–31.

Thompson, R. 1995. Complex demodulation and the estimation of the changing continentality of Europe's climate. *International Journal of Climatology* 15: 175–85.

Tsonis, A. A., and Elsner, J. B. 1988. The weather attractor over very short time-scales. *Nature* 333: 545–7.

Vandaele, W. 1983. *Applied Time Series and Box–Jenkins Models*. New York: Academic Press.

Velleman, P. F., and Hoaglin, D. C. 1981. *Applications, Basics and Computing of Exploratory Data Analysis*. Boston, MA: Duxbury Press.

Wigley, T. M. L. 1986. Commentary. In Clark, W. C., and Munn, R. E. (eds) *Sustainable Development of the Biosphere*. Cambridge: Cambridge University Press, 289–91.

Wilcox, B. P., Seyfried, M. S., and Matison, T. H. 1991. Searching for chaotic dynamics in snowmelt runoff. *Water Resources Research* 27: 1005–10.

5

FUTURE GLOBAL WARMING

Resolving the climatologist and economist conflict

Max Wallis

Editors' note This chapter is the first in the book to have a specifically 'future' orientation. There are several key points that emerge from it. First there is a technical argument over whether we should assess future impacts by using a cut-off time horizon – say we calculate the costs of warming over a 100-year time-scale as the climatologists usually do – or by using a discounting approach which can consider an infinite future – the economists' usual approach. Then there is a discussion of the calculation of costs, on the one hand by considering a particular change in mean temperatures, and on the other by considering not the absolute change, but the rate of change instead. Max Wallis concludes that this is probably the more significant indicator: we have to keep rates of change sufficiently low that ecosystems and society can have a realistic hope of adaptation. And in this case the discounting approach is better than the time horizon approach. In Chapter 10 ironically Malte Faber and John Proops introduce the idea of a time horizon cut-off for an entrepreneur – arguing that that is how economic individuals do indeed behave. The problems of rate of change and of adaptation have been considered by Jean Grove in a historical context in Chapter 3. This chapter is also a very useful introduction to the complexities and disagreements behind the scientific assessment of greenhouse warming with which the politicians grapple in Ros Taplin's following chapter, Chapter 6. Wallis also begins to touch on the incapacity of the international system of sovereign states and market economies to handle the grave crisis he sees, but is unable within this chapter to pursue that new line of enquiry. We do not agree with his view that it is 'irresponsible' to argue that developing countries should be allowed greenhouse gas emissions as great as those of the industrialised countries in the past. It is not an argument, but a fact, that under the current system of states they will simply do what they see as necessary to relieve poverty – and the best hope of achieving that is to help in an economic transition on as short a time-scale as possible, since that will also accelerate the stabilisation of population levels. We have noted above that the poorest are most at risk from change, having fewest reserves to cushion the impact of change, and that is why poor countries reserve the right to increase the wealth of their people by what they see as the most appropriate means. To persuade them to do otherwise will require concessional technology transfer from the industrialised nations.

INTRODUCTION

Of all the greenhouse gases, CO_2 is exceptional in that its emission affects the atmosphere for geological time-scales. Much of the CO_2 emitted by fossil fuel burning dissolves in the oceans within 100 years or so, but a fraction remains in the atmosphere for geological times, as ultimate deposition in ocean sediments is very slow. The greenhouse gas methane, on the other hand, reacts chemically in the atmosphere on a 10-year time-scale, producing ozone (a short-lived but very strong greenhouse gas), water vapour (significant only when produced high in the stratosphere), and CO_2. The impact I of a pulse of gas on the global climate is gauged by integrating its radiative strength (termed 'forcing') over time, averaged through the whole atmosphere (Wigley 1994):

$$I = \int WRq(t)\mathrm{d}t \qquad (5.1)$$

where W is a weighting factor, R is the radiative strength, and $q(t)$ denotes the fraction of gas remaining at time t after emission.

Figure 5.1 gives examples of q for pulses emitted at time zero. Most greenhouse gases decay in the atmosphere as a simple exponential of time; a pulse of nitrous oxide (N_2O) decays in hundreds of years, whereas a pulse of methane (CH_4) decays in decades. A pulse of CO_2 behaves, as Figure 5.1 shows, in a more complex fashion, relating to somewhat uncertain behaviour of the ocean systems and to levels of CO_2 in the future atmosphere. The 1990–2 standard model (IPCC 1990, 1992) is shown as the broken CO_2 line; the 1994 reference model (heavy line – IPCC 1995) decays much faster in the first 100 years. The latter assumes that future CO_2 levels are stable, whereas a CO_2 pulse persists more strongly under increasing CO_2 levels (broken heavy line). The 50 per cent uncertainty shown as differences between the CO_2 models translates into uncertainty in CO_2's integrated impact I. Note that the profile of CO_2 does not have a simple decay constant, nor single decay time.

Figure 5.1 shows that, unlike N_2O, a pulse of CO_2 stays at a significant level of 20–35 per cent after a few centuries; at least, this is thought to be reasonably true, though applicability of the model becomes increasingly uncertain centuries hence. Because of this persistence, the integral given in equation (5.1) is calculated from the time of emission $t = 0$ up to a 'time horizon', T, commonly taken to be 100 years. Conventional 'global warming

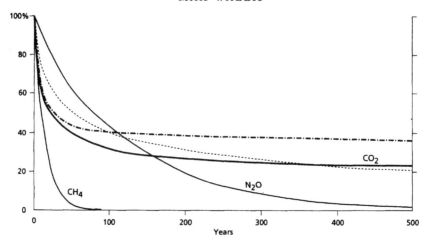

Figure 5.1 Decay with time (years) of pulses of greenhouse gases after emission to the atmosphere. For CH_4 (methane) and N_2O (nitrous oxide) the decay is exponential. For CO_2 the broken curve was formerly used by IPCC (1990, 1992), the continuous curve now favoured by the IPCC (1995) under stabilised greenhouse gases, and the upper dot-dash curve is applicable if greenhouse gases continue to increase. The slowly diminishing tails of the CO_2 models illustrate the point that some fraction of the initial impulse (15–25 per cent) remains in the atmosphere for indefinite time (according to the current carbon cycle model)

potentials' for the various greenhouse gases are then defined as the ratio $GWP[gas] = I[gas] : I[CO_2]$.

Some climatologists have adopted much longer times T, 1,000 or 3,000 years (Lashof and Ahuja 1990; Edmonds and Wuebbles 1988). But that does not escape the absolute time problem – that in reality today's CO_2 emissions are predicted to pollute the global environment for hundreds or thousands of human generations.[1] Ethically, the problem is similar to the equitable sharing of a finite resource between indefinitely numerous future generations (Dasgupta and Heal 1979; Broome 1992). Thus, formulating the time horizon measure of CO_2 implies acquiescence in the view that progress will take care of posterity – an attitude that conflicts with sustainability. Climatologists often express distrust of economists for 'discounting' the future, but their own use of time horizons does much the same. The economists' up-front approach takes W to decay exponentially as $\exp(-rt)$, with r the discount rate. Figure 5.2 shows the economists' W in the case $r = 1/T$, compared with the discontinuous W of the climatologists; the two functions have equal area. Under the economists' function, the 20 per cent or so CO_2 that persists for millennia contributes in this case little to the integral (5.1), which may as well be extended to 'infinite' time. It is now accepted that the integrated radiative forcing I of (5.1) is not purely geophysical, but is a user-oriented construct that involves both some earth-system processes and some value choices

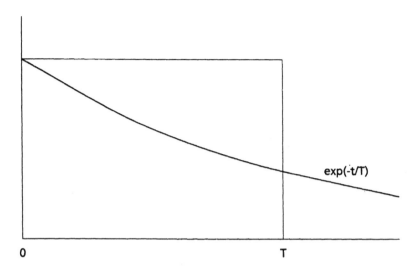

Figure 5.2 The two weighting functions *W* changing with time: *T* is the time horizon for the climatologists and the decay time or inverse discount rate that for the economists

(Wallis and Lucas 1994; IPCC 1995, sect. 5.1.2). There is no way, as Smith and Ahuja (1990) have stressed, to avoid allocating weights that effectively count impacts differently for present and future human generations. And how one does this affects the relative *I*-values of the various greenhouse gases calculated by equation (5.1). The IPCC consensus-oriented process tended initially to de-emphasise this critical point and keep the real scientific and socio-economic uncertainties out of the public domain. While the IPCC professed to take no stand, its adoption of time-limited GWPs and its homing in on results for the 100-year time horizon (IPCC 1990, 1992) have encouraged too narrow an approach and misleadingly precise quantification.

If economists wish to calculate the harm (or benefit) of the greenhouse gas, they could call *W* a 'cost' discounted to present value, but probably including some increase with time, to allow for higher marginal cost relative to the accumulated changes in the atmosphere. An alternative formulation is to take all gases as proportional to CO_2 and choose *W* so as to convert *I* into a temperature increase θ. Economic damage and remediation costs are supposed to increase faster than in direct proportion to CO_2 increase, perhaps as fast as the square of CO_2 increase (Laurmann 1980; Nordhaus 1993).

TIME-SCALES FOR IMPACTS AND CONTROL MEASURES

Undesirable or risky climatic effects would occur on diverse time-scales (WMO 1991). There are changes, such as rise in sea-level or decreases in regional rainfall, that can inundate or desertify significant areas. Rapid climate change over successive decades is adverse in so far as climate zones shift more rapidly than natural systems can adapt.

According to the WMO report (1991), the choice of T is left to the analyst or policymaker, depending on which type of undesirable change most interests her. The Rio Framework Convention on Climate Change moved away from such subjective choice to considerations of adaptation and economics, formulating the objective 'within a time frame sufficient to allow ecosystems to adapt naturally to climate change, to ensure that food production is not threatened and to enable economic development to proceed in a sustainable manner'. Note, however, that the indicators in Table 5.1 do not appear to cover the extreme events that are the most damaging: hurricane-force storms, tidal storm surges, extended drought or floods.

The concept of adaptation in the Rio Framework Convention is closely related to time-scale and rate of change. Both ecosystems and human societies adapt so as to tolerate increased swings in temperature, changed rainfall, and severer climatic extremes. Ecosystems have adapted to gradual climatic changes throughout global history; indeed, climate variability is one of the driving forces of evolutionary change and biological diversification. Human society would adapt by changing agricultural practice, resettling away from areas vulnerable to flooding, or planting new crops. These adaptive measures require substantial new investment and the writing off of old ones. Similarly, technologies for combating growth of greenhouse gases – termed mitigation measures – require investment in new processes and the writing off of the old. Fossil fuel systems are being designed that extract CO_2 and dump it in old gas fields or in the deep ocean. Afforestation can also remove CO_2; an area equal to 5 per cent of present forests added by 2020 is a practicable target (Kelly 1990), with the timber in which the carbon is locked up being used as building material. Methods for combusting or oxidising methane from landfill sites are being devised, but we may have to stop landfilling biodegradable material and biodigest it or compost it in

Table 5.1 Climatic change indicators and their integration time

Climate change indicator	Integration time, T
Change in average temperature <θ>	~100 yrs
Rate of change of temperature θ	20–50 yrs
Change in mean sea-level	>100 yrs
Rate of change of sea-level	>50 yrs

controlled ways. Leakage of methane in natural gas production and distribution can be curbed, and a massive programme of renewing old gas mains (60 per cent of UK distribution mains are pre-1969 jointed metal) may be required. The hydrochlorofluorocarbons (HCFCs), replacing chlorofluorocarbons (CFCs) in some applications, are very strong greenhouse gases so may have to be withdrawn. Power technologies that are CO_2-free, especially wind, wave, and biomass, can replace fossil fuels.

There is thus a wide range of potential technologies for mitigating the greenhouse effect; they generally carry an economic penalty, if only in development and capital costs. The safeguarding of food production and sustainable economic development, under the Rio objective, also points to economic valuation. The costs of damage via climate change are to be compared with the costs of mitigation measures, including reducing emissions. This can be formulated as a classic investment problem, with time-scale inherent in the canonical socio-economic rate of return (3–4 per cent/yr). The choice of discount rate is therefore subject to the considerations which surround the choice of the same parameter in normal investment. The appropriate choice is the rate for risk-free investment (Pearce et al. 1989), but modified by a risk-averse strategy. This use of discounting is evidently justified (Norgaard and Howarth 1991) as guiding the efficient use of our generation's resources. Some analyses relate the discount rate to a social time-preference (survey of Lind 1982), taken as 3 per cent/yr (Nordhaus 1993), or less if one is uncertain about discounting over the long term (Fankhauser 1994), but this rate still depends on economic success continuing more or less steadily into the future.

Many analysts argue that uncertainties in future global climate change are substantial: Schneider (1993) has conceived various prospective surprises, from superhurricanes to political instability. There is substantial risk to the entire global economy (and environment), not to a small subsection. It is to be hoped that avoidance measures (greenhouse gas mitigation) will be adopted, collectively and at various future times dependent on then current data and theoretical prediction. The damage and risk are related in complex ways to knowledge and geopolitics as well as time. The large uncertainty and risk is sufficient reason to challenge the neoclassical economic model (Dowlatabadi and Lave 1993). The 'precautionary principle' requires human society to steer well clear of uncertain thresholds of severe damage (FoE 1990). Indeed, the large 'costs' of worst-case outcomes have been shown via decision-analytic theory (Laurmann 1980) to condition a risk-averse strategy;[2] according to these results, this means acting as if a critical level (e.g. CO_2 doubling) is expected some 15–20 years earlier than under the mean climate prediction. Aggressive measures to keep greenhouse gases well below potentially catastrophic levels are to be favoured (Cline 1992). And inclusion of delays in changing to new technologies (Laurmann 1980) would justify measures to avert the risk as early as the present day.

IMPACTS DEPENDING ON RATE OF CLIMATIC CHANGE

Most attention has been given to the favourite indicator: the global average surface temperature $<\theta>$. But in some regions of the earth, annual rainfall or length of drought periods are more important (Roberts, Chapter 2). Moreover, certain effects of atmospheric change depend more on the rate of change than on the change in itself (Table 5.1). Low rates of change are tolerable, indeed are accommodated naturally, but severe damage results from high rates of change. While human societies and ecologies naturally renew and evolve, rapid climate changes can be too speedy for normal economic renewal or for species migration and acclimatisation; high costs for infrastructure renewal or impoverishment of ecological systems can result. The world's oceans are a source of inertia in the climate system, tending to delay climatic response and consequent impacts. And as the oceans adapt less rapidly than the land to changes in atmospheric forcing, so the stronger land-to-sea temperature differences may well produce stormier weather. Such qualitative arguments have led to proposals to replace q in equation (5.1) by its rate of change, q' (Wallis 1994), or analogously replace θ by θ' (Peck and Teisberg 1992). These alternative measures of impact do not encounter the time horizon problem, but they do exclude the long-term effects of permanently raised CO_2 (e.g. loss of coastal land to raised sea levels).

We can readily compare the two approaches in the simplest case of a single gas which depletes with decay constant and 'costs' C proportional to the greenhouse effect g or its rate of change g'. Algebraic formulae derived (Wallis and Lucas 1994) from an adaptation of (5.1) show total costs into the future (Table 5.2). In Table 5.2, T is the time horizon, r the discount rate and α the gas decay rate as in Figure 5.1. Though algebraically different, the expressions in the first column for the two cases are much the same when r is read as $1/T$; those in the second column are, however, very different for smaller values of r and correspondingly large T. The time horizon case gives a poor result because of the artificial cut-off at time T. This is evident in Figure 5.3: the upper figure shows the two functions from the left-hand column differ little, while the lower figure shows increasingly large relative differences to the right (large T).

Thus 'global warming potentials' (GWPs) on the usual definition, proportional to the instantaneous atmospheric gas as in equation (5.1), are

Table 5.2 Costs of greenhouse change under time horizon and discounted formulae

	Cost ~ greenhouse effect $\times R \times (\partial C/\partial g)$	Cost ~ rate of change $\times R \times (\partial C/\partial g)$
Time-horizon case	$(1 - e^{-\alpha T})/\alpha T$	$(e^{-\alpha T})/T$
Discounted case	$r/(r+\alpha)$	$r^2/(r+\alpha)$

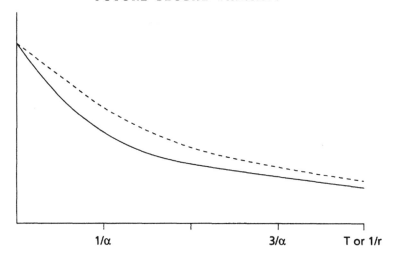

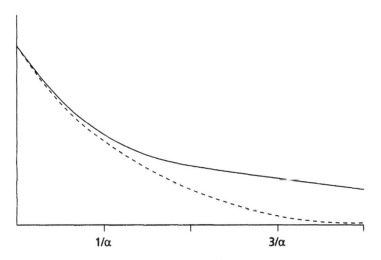

Figure 5.3 Representations of the global warming potential functions of Table 5.2.
The upper curve shows the ordinary GWPs (left-hand column of the table) and the
lower curve shows rate-of-change GWPs (right-hand column) for values of time
horizon T (broken line) of discount rate r (solid line)

little affected by the way time is handled. A time horizon approach is
equivalent to discounting for evaluating 'ordinary' GWPs (first column of
Table 5.2), and the appropriate T is roughly the inverse discount rate, $1/r$. But
if GWPs are defined as being dependent on the rate of change (second
column), only the discounting approach gives a reasonable measure. For
CO_2, the assessment is a little more complex, as this gas behaves as a sum of

three or more component gases of differing decay constants (see Figure 5.1). It is again better to use the discounting approach so that the arbitrary 1,000 or 3,000-year cut-off is avoided.

Threshold arguments also lead to preference for indices relating to the rate of increase of greenhouse gases. Global and regional climate changes occur without anthropogenic intervention and, indeed, are drivers of evolution and ecological diversity. Studies of past climates – palaeoclimatology – from the records in sediments, fossils, and Antarctic ice layers show how the climate and greenhouse gases (CO_2 and methane) have changed over past millennia, and provide insight into climate mechanisms and ecological responses. Since we emerged from the last Ice Age 15,000 to 10,000 years ago, the climate has warmed by an average of 0.25°C per century, and various species including forest trees slowly migrate following the deglaciation, at some 10–100km per century (Huntley 1990). It is argued that this migration rate was the fastest achievable naturally. Ecological systems comprise a range of species, including some less mobile and some more tolerant of stress. So ecological impoverishment – a less diverse and less robust ecosystem – probably results for higher migration rates. On a regional scale, emergence from the last Ice Age has been at times faster than 0.25°C per century, and experts have fastened on 1°C per century as the tolerance level of forest communities (Kelly 1990). (In comparison, present global warming is judged to be about 2–4°C per century (IPCC 1992).) The idea, then, of natural adaptability to natural climate changes suggests there is a threshold of sustainable rate of change, which is currently expressed as a rate of global mean temperature rise of roughly 0.1°C per decade.[3]

Stabilisation of atmospheric concentrations of greenhouse gases is sometimes given as a target, based on the IPCC (1990) tabulation of emission reductions required to stabilise at 1990 levels (IPCC 1990, Policy Makers Summary). Not only is this impracticably far out as a target (over 60 per cent cuts in CO_2 stated), but it also ignores the importance of rate of change. A short-lived gas such as methane has emission rates automatically close to stabilisation at 1990 concentration, but there is no justification for accepting the 1990 level at over twice the pre-industrial level as a target. Costs of reduction below the 1990 level may be relatively cheap, at least for mitigating methane from industrial processes and landfill sites.

Rate-of-change and threshold approaches also help bypass the irresponsible argument that developing countries should be allowed high emissions of CO_2, etc., comparable to the large totals emitted by the developed countries in the past. The climate threat faces the global community as a whole, and arises from the unsustainable rate of change of greenhouse gas levels. It arises more from the future emissions than from past emissions, includes agriculture and deforestation impacts, and covers the range of degradable non-CO_2 gases. The sustainable rate of change concept implies a focus on ways to share equitably the corresponding allocation of greenhouse gas emissions.

DISCOUNTING THE FUTURE

Implicit in the choice of time horizon GWPs are certain value judgements that give the GWP function social relevance, hence putting it into the ambit of social science. The time horizon approach, using a weighting constant up to time horizon T and zero thereafter, has led to a quagmire of subjectivity. The alternative discounted weighting is an effective substitute and is more powerful in encompassing rate-of-change dependency. The IPCC update (IPCC 1992) specified various limitations of GWPs: primarily inadequate understanding of CO_2 uptake on the global scale (the 'carbon sink' problem is unresolved); indirect forcing, e.g. via methane, generation of ozone is quantitatively uncertain; and averaging over regional, seasonal, and altitude variations is necessary for some components and thus model dependent. These are obviously practical limitations, rather than principled ones. Nevertheless, the IPCC still declared, without justifying the statement, that values of T equal to 20-, 100-, and 500-yr periods are 'believed' to provide a practical range. But where discounting is appropriate, for investment in mitigation measures and social adaptation, $T = 500$ yr and $T = 100$ yr appear quite unrealistic. The 3–4 per cent social discount rate is approximately equivalent to a 30-year time horizon, considerably shorter than commonly adopted for GWP assessments. In practical terms this implies that measures to reduce emissions of short-lived gases such as methane (15-yr lifetime – IPCC 1995) rise in importance.

In so far as the impacts on individuals and societies of climate change can be given economic values, cost–benefit analysis would apply, with weighting by a discounting function, i.e. exponential analysis instead of regular 'time horizon' weighting. Subjective choices are not avoided – selecting W requires assessment of impacts on populations and economics – but choices can be aligned with established socio-economic practice. It is argued that the big uncertainties and high cost risks strongly change conclusions of cost benefit analysis, in the direction of aggressive measures to keep damage well below catastrophic levels (Cline 1992).

Discounting of mitigation and adaptation costs falling on generations centuries hence is, however, dubious. There is no guarantee that economies will continue to grow. Investments to deal with pollution from our generation, such as radioactive waste, may be uncashable by future generations. Some argue that long-term damage is acceptable, on condition of establishing a specific compensation mechanism (Pearce et al. 1989). Present models in which steady economic growth will provide the funds for remedying greenhouse impacts (Nordhaus 1992 and 1993) have no such mechanism. Nor is any shown to be viable. Climate changes will fall inequitably on states and communities, so the mechanism needs to be globally redistributive and thus is not readily compatible with independent states and market economies. This is surely a fundamental problem with neoclassical economic models.

The alternative view distinguishes the issue of resource rights for future generations, for which discounting is not appropriate (Norgaard and Howarth 1991). We hand on to future generations not, primarily, short-lived manufacturing capacity, but natural resources, including the holistic global resource. This implies sustainability constraints (FoE 1990), fitting in with the Brundtland definition of sustainable development and the concept of global commons. In the case of the global climate system, it means handing on not a static system with stabilised atmospheric gases, but a system changing tolerably slowly, robust to uncertainties and far separated from potentially catastrophic changes – conditions that surely insist, under the present state of knowledge, that we take stringent measures to limit the release of greenhouse gases.

AFTERWORD

For my straying from physical science into economics, I owe the initial stimulus to Jonathon Porritt – to his emphasis that environmentalists who want to change the world have to get into economics. As an atmospheric scientist too, I have remained critical of the IPCC focus on 'consensus' and so-called best estimates.

The scientist's tacit justification – that politicians cannot handle uncertain predictions – is not wholly true, and I believe that public confidence in the science has suffered through evident changes in the successive IPCC assessments (1990, 1992, 1995).

NOTES

1   Carbon is slowly sequestered in corals and new sedimentary rocks, but it is not clear that increased atmospheric CO_2 will increase the sequestration rate. Moreover, the global climate changes naturally, so an interglacial period, as discussed in Chapters 2 and 3, may be the relevant comparator, say 3,000 human generations.

2   Nordhaus (1993) uses a 'risk aversion' parameter as a multiplier of the rate of income growth (see Lind 1982), to allow for increased monetary damage to a future economy. Laurmann (1980) uses 'risk averse' with its wider meaning, assuming an exponential utility function to strongly weight damage costs when approaching some fraction of the net global product.

3   On regional scales, much faster changes have occurred without global catastrophe, again as discussed in Chapters 1 and 2. Such faster changes result from changes in ocean currents, changes in atmospheric circulation patterns, and changes in vegetation cover. This may indicate that $<\theta>$ and θ' are not good parameters of global change, but does not prove that global climate and ecology are robust to 'external' forcing by greenhouse gases.

REFERENCES

Broome, J. 1992. *Counting the Cost of Global Warming*. Cambridge: The White Horse Press.

Cline, W. R. 1992. *The Economics of Global Warming*. Washington, DC: Institute of International Economics.

Dasgupta, P. S., and Heal, G. M. 1979. *Economic Theory and Exhaustible Resources*. Cambridge: Cambridge University Press, 255–82.

Dowlatabadi, H., and Lave, L.B. 1993. Letter: Pondering greenhouse policy. *Science* 259: 1382.

Edmonds, J. and Wuebbles, D. 1988. *A Primer on Greenhouse Gases*. Washington, DC: US Department of Energy, NBB-0083.

FoE (Friends of the Earth) 1990. *Beyond Rhetoric: An Economic Framework for Environmental Policy Development in the 1990's*, ed. Gee, D. *et al*. London: FoE.

Fankhauser, S. 1994. Evaluating the social costs of greenhouse gas emissions. *Energy Journal* 15(2): 158–84.

Huntley, B. 1990. Lessons from climates of the past. In Leggett, J. (ed.) *Global Warming: The Greenpeace Report*. Oxford: Oxford University Press, 133–48.

IPCC (International Panel on Climate Change). 1990. *Climate Change: The IPCC Scientific Assessment*, ed. Houghton, J. T., Jenkins, G. J. and Ephraums, J. J., Cambridge: Cambridge University Press.

IPCC (International Panel on Climate Change). 1992. *Climate Change: The IPCC Scientific Assessment 1992*, ed. Houghton, J. T., Callender, B. A. and Varney, S. K., Cambridge: Cambridge University Press.

IPCC (International Panel on Climate Change). 1995. *Climate Change: 1994 Radiative Forcing of Climate Change*, ed. Houghton, J. T., and Filho, M., Cambridge: Cambridge University Press.

Kelly, M. 1990. Halting global warming. In Leggett, J. (ed.) *Global Warming: The Greenpeace Report*. Oxford: Oxford University Press, 83–112.

Lashof, D. A., and Ahuja, D. R. 1990. Relative contributions of greenhouse gas emissions to global warming. *Nature* 344: 529–31.

Laurmann, J. A. 1980. Assessing the importance of CO_2-induced climatic changes using risk-benefit analysis. In Bach, W. *et al*. (eds) *Interactions of Energy and Climate*. Dordrecht: Reidel.

Lind, R. C. 1982. *Discounting for Time and Risk in Energy Policy*. Washington, DC: Resources for the Future.

Nordhaus, W. D. 1992. An optimal transition path for controlling greenhouse gases. *Science* 258: 1315–19.

Nordhaus, W. D. 1993. Optimal greenhouse gas reduction and tax policy in the DICE model. *American Economic Review* 83: 313–17.

Norgaard, R. B., and Howarth, R. B. 1991. Sustainability and discounting the future. In Constanza, R. (ed.) *Ecological Economics: The Science and Management of Sustainability*. New York: Columbia University Press, 88–101.

Pearce, D., Markandya, A., and Barbier, E. B. 1989. *Blueprint for a Green Economy*. London: Earthscan.

Peck, S. C., and Teisberg, T. J. 1992. Temperature change related warming cost functions: a further analysis with CETA. Preprint, EPRI, Palo Alto, CA.

Schneider, S. H. 1993. Pondering greenhouse policy. *Science* 1381: 259.

Smith, K. R. and Ahuja, D. R. 1990. Towards a greenhouse equivalence index: the total exposure analogy – an editorial. *Climatic Change* 17(1): 1–7.

Wallis, M. K. 1994. Greenhouse coefficients dependent on rates-of-change. In van Ham, J. *et al*. (eds) *Non-CO_2 Gases*. Kluwer.

Wallis, M. K., and Lucas, N. J. D. 1994. Economic global warming potentials.

International Journal of Energy Research 18: 57–62.

Wigley, T. M. L. 1994. The contribution from emissions of different gases to the enhanced greenhouse effect. In Hanisch, T. (ed.) *The Rio Convention on Climate Change: The New Regime and Agenda for Research*. Boulder, Colo.: Westview Press.

WMO (World Meteorological Organization). 1992. *Scientific Assessment of Ozone Depletion* 1992. Geneva: WMO Global Ozone Res. and Mon. Project.

6

GREENHOUSE POLICY DEVELOPMENT AND THE INFLUENCE OF THE CLIMATE CHANGE PREDICTION TIMETABLE

An Australian perspective

Ros Taplin

Editors' note This chapter introduces us to the manner in which the political process confronts the perceived threat of global warming. What it does not fully establish is why a long-term and uncertain threat has been accepted as sufficiently real to warrant attention within the short-term decision-making process of democratic politics, although there are some clues given. It points out that the political process at both international and national levels is actually running faster than increases in scientific knowledge – in other words, that the political process, once started, has its own momentum, at least for a while. In an address in London in 1994 Sir John Mason, an eminent atmospheric scientist closely connected with the development of global circulation models, actually argued that the scientists now knew that global warming would not occur as fast as originally thought, and told the political community to reduce its anxieties and activities for the next decade, while giving the scientists huge sums of money to improve their knowledge of the deep oceans. Part of the answer to the riddle may be that the public, in Australia in this chapter, are concerned over green issues – and of course urban areas have their quota of local atmospheric pollution. This the politicians know, and at the international level, as Ros Taplin points out, the greenhouse issue can symbolise the interconnectedness of the global ecosystem. There is no doubt that the international community is competitive in terms of agenda setting and the gathering of Brownie points, as pointed out in Chapter 1 in the case of Rostow's 'take-off' and the UN Development Decade. Now, post-Brundtland, instead of 'fast forward', we have as 'visions of the world or the future' (Godard, quoted by Taplin below) the concepts of 'sustainable development' and 'hold-fast' (stabilise emissions). The Brownie points can be gathered now by entering into rather vague and long-term commitments. When it comes to working out a strategy for local implementation, however, short-term and much more certain realities tend to prevail: negative economic impacts such as the closing of certain industries

121

are not allowed. Taplin illustrates all these points and many more in showing how after the rhetoric the Australian Strategy evolved into a plan of little change. The chapter is complemented neatly by John Gordon's following chapter, which concentrates mostly on a similar pattern of events in the UK.

INTRODUCTION
Greenhouse time-scale and policymaking implications

The climatic changes predicted for the next fifty to one hundred years as a result of anthropogenic greenhouse gas emission are relatively minor in terms of the earth's climatic history. Far more severe events, namely ice ages, have occurred relatively recently in the earth's geological time-scale, with the last one ending about 10,000 years ago. However, when greenhouse predictions are viewed in terms of the time-scale of individual human life, the environmental change predicted is very much linked with its potential to dramatically affect human existence in the short term; that is, in the order of decades. The concept of environmental change linked with the greenhouse effect is thus very much linked to the span of a human life or the time-scale of a human generation.

Predictions of the impacts of climate change are based on global climatic model output data on future atmospheric concentrations of greenhouse gases.[1] The World Meteorological Organization–United Nations Environment Programme body, the Intergovernmental Panel on Climate Change (IPCC), updated scientific assessment for its 1992 reports for a 'business-as-usual' scenario, a CO_2 doubling time of 60 to 100 years[2] and a globally averaged mean increase surface air temperature of 1.5–4.5°C (IPCC 1992b). Bert Bolin, IPCC chair, has suggested that such a greenhouse gas buildup may lead to significant climate change impacts in the next twenty to thirty years; that is, in a generation's time. Bolin also has said, 'It is clear from IPCC assessments that climate change cannot be avoided unless emissions of CO_2 are reduced' (Bolin 1994). Countries with environments at risk according to the Framework Convention on Climate Change include small island countries, countries with low-lying coastal areas, countries with arid and semi-arid areas, countries with areas prone to natural disasters, countries liable to drought and desertification, countries with severe urban atmospheric pollution, and countries with areas with fragile ecosystems.

Even though climate change predictions focus on the time-scale of only a few decades ahead – less than the planned life of a city high-rise – greenhouse-induced climate change has emerged to challenge fundamentally the way politicians and decision-makers view the time-scale of environmental problems. Climate change differs from most other environmental problems dealt with by policymakers in that it is an issue that is long term in comparison to political time-scales where the short term is one or two years and the long term is arguably ten years. As Christie (1992) has observed, 'The timetable of elections and the personal career timetable of individual politicians are inevitably at odds with the timeframe of global warming.' Environmental problems of the type that politicians normally are called on to address are those of the present or immediate future: for example, air pollution in cities, logging of forests or even ozone depletion. Graham Richardson, a former Australian senator and Minister for the Environment, said somewhat optimistically in 1989 at the height of public concern in Australia about greenhouse warming:

> what greenhouse is going to do to [Australian] politics is that it will change – or at least it ought to be changing – politicians from looking at the next three years to the next election. We should be looking one or two generations ahead. I hope that this may be the one issue that goes past ordinary politics and rewrites our political history books. Perhaps it will be the one issue that will give us a real opportunity for change.
>
> (Richardson 1990 pp. 69–70)

As well as the unusual temporal dimensions of environmental change due to the greenhouse effect, the nature of the spatial dimensions of predicted climate change impacts is unusual: the sources or causes of the impacts of climate change on socio-environmental systems arise in some cases very far away from the local situation. The impact of the emissions from the industrialised North on the developing nations in the island South Pacific is a case in point. These small island states are highly vulnerable to the impacts of climate variability, which is manifest in extreme events such as tropical cyclones. Accordingly, scientific warnings about the possibility of current climate extremes becoming more severe in the future owing to the greenhouse effect have been taken most seriously by these states (Taplin 1994).

Climate change thus has global causes and global consequences, and requires a coordinated global community response in order to deal with its predicted impacts. Climate change policymaking belongs to the category of large-scale policy and is 'unconventional' or 'supra-political' (Schulman 1980); it has been referred to as an 'extraordinary societal undertaking' (Rhodes 1992).

Arguably, among the policymaking community the greenhouse effect has fostered an acknowledgement of the global interconnectedness of environmental problems, emphasising the need for sustainable development as a long-term ecological goal, and for environmental policymaking that incorporates the precautionary principle and intergenerational equity. Attempts to respond to greenhouse predictions through public policymaking are being undertaken both unilaterally and multilaterally by many developing and developed nations; the negotiation of the UN Framework Convention on

Climate Change, which came into force on 26 March 1994, has been driven by international concern.[3]

Uncertainty about the greenhouse effect

As indicated above, greenhouse policymaking is a response to predictions of environmental change well into the twenty-first century. Knowledge of the time-scale of the impacts of climatic change and the degree of severity of their impacts at the regional scale[4] is uncertain. None the less, scientific concern about the greenhouse effect has spurred policymakers internationally to address seriously the issue even in the face of scientific uncertainty. According to the IPCC science working group, 'the unequivocal detection of the enhanced greenhouse effect is not likely for a decade or more' (IPCC 1992b). However, the hundreds of leading scientists involved in the IPCC assessment undertaking are arguably involved because of their concerns that policy action should be taken to avert the potential impacts of climatic change. As Schneider (1991) has said, 'The public policy dilemma is how to act even though we will not know in detail what will happen.... Public policymakers will have to address how much information is "enough" to warrant action ... we will have to rely on the intuition of experts' (p. 53).' The time-scale of greenhouse predictions indicates that decision-makers need to act now to mitigate emissions of greenhouse gases and to put into place adaptation policies, but greenhouse impacts are in the time-scale of decades. Bolin has said recently,

> The risk is there. It is not certain that there is a major [hu]man-induced change of climate, but there is a risk that it might happen. Personally, I think that there will be a change. The question is rather: how quickly it will occur, what it will look like, and how serious will it be.
>
> (van Zijst 1994, p. 4)

Uncertainty, then, highlights a further time phenomenon with regard to the greenhouse effect: that is, political moves to develop policy, at least symbolically (Edelman 1964), at both international and domestic levels, are running ahead of developments in scientific knowledge about climate change. The existence of the Framework Convention on Climate Change testifies to this and exemplifies the particular type of environmental decision-making process that Godard (1992) describes as evolving 'in controversial contexts, [with] an association of ignorance, uncertainty, potential gravity and irreversibility'.

Factors in the Australian perspective on greenhouse policymaking

The remainder of this chapter explores Australian greenhouse policymaking and discusses the influence of the climate change prediction timetable. First, some background is given on Australia's role in international environmental policymaking and in the development of the international climate change regime; second, a brief profile of Australian public opinion on environmental issues and in particular on greenhouse warming is given; third, Australia's domestic climate-change policy response and the factors shaping it, including the time-scale of the greenhouse issue, are discussed; and, finally, conclusions are drawn about the political, economic, and time-scale dimensions of the climate change issue.

AUSTRALIA'S ROLE IN INTERNATIONAL ENVIRONMENTAL POLICYMAKING

Over the past two decades, Australia has taken a close interest in and has worked in cooperation with other nations on many environmental problems.[5] Australia's environmental profile internationally has been referred to as one of 'enlightened self interest' (Evans and Grant 1991), 'a nation with an environmental conscience' (Henderson 1992), and as 'playing constructive roles as a good international citizen' (Boardman 1991). This at least is how Australians would like to be seen abroad, even if in this chapter no evidence is presented on how that image is actually received by Asian neighbours and other countries. Certainly, the decision in 1989 to create a position of Ambassador for the Environment as a focus person for Australia's international environmental activity was motivated to some degree by a perception that Australia should play a strong and effective role in international environmental diplomacy. This behaviour is in accordance with Godard's (1992) observation that 'competition between countries ... now also involves involves "visions of the world or the future" (i.e. explanations of environmental phenomena and scenarios of economic development)'; he says these are 'products of complex mixes of individual beliefs, social rules, policies, administrative regulations and institutions'. Ian McPhail, Director of Australia's Commonwealth Environment Protection Agency, has said about Australia's role in international environmental policymaking:

> Australia has developed a reputation of being the 'honest broker'. We are seen as being able to liaise effectively with both developed and developing countries. This may be because of our strong commitment to multinationalism and to international consensus building. We have been progressive and forward-thinking in areas of sustainable development; and are recognised as the only developed country with a 'mega-diverse' biology. Our stand on mining in Antarctica has also provided us with a good reputation internationally.
>
> (McPhail 1992)

Australia's commitment towards protection of the global environment has also been evident in the adoption by the Australian government of the

Toronto target[6] as a national interim planning target for greenhouse gas emission reduction. The policy is to stabilise greenhouse gas emissions, based on 1988 levels, by the year 2000 and to reduce these emissions by 20 per cent by the year 2005. However, it should be noted that economic and trade caveats were attached to this target; it is 'subject to Australia not implementing response measures that would have net adverse economic impacts nationally or on Australia's trade competitiveness, in the absence of similar action by major greenhouse gas producing countries' (Australian government, Intergovernmental Agreement on the Environment, 1 May 1992). The interim target decision was made in October 1990 in preparation for the Second World Climate Conference. Ros Kelly (then Australian Environment Minister) said to the magazine *Bulletin* in reference to the decision that she 'wanted an Australian commitment to a national reduction of 20% in greenhouse gas emissions so that Australia would be one of [the] ... nations at ... [the] conference in setting an example for the world' (Barnett 1990). Of the thirteen nations represented at the Second World Climate Conference in Geneva in October–November 1990 (Jager and Ferguson 1991), Australia's emission reduction target was one of the most stringent (Zillman 1990).

In playing an active role in the international climate change regime formation process, Australia has actively participated in the Intergovernmental Negotiating Committee for the Framework Convention on Climate Change and was the eighth nation to ratify the Convention.[7] Also, Australian scientists and bureaucrats have contributed considerably to the science, impacts, and response strategies work of the IPCC. Researchers from the Bureau of Meteorology, the CSIRO (Commonwealth [of Australia] Scientific and Industrial Research Organization), and Australian universities, together with officials from federal government departments (Foreign Affairs and Trade; Environment, Sport and Territories; and Primary Industries and Energy), have invested a substantial amount of time in the IPCC process.

Australia's contribution to the greenhouse regime, as well as stemming from a desire to assume environmental leadership, or at the least the role of a good international environmental citizen, also comes from economic self-interest. The potential implications of reducing greenhouse gas emissions for Australia as an energy and energy-intensive products exporter, and especially because of the major importance of coal, are a very significant economic concern. A converse economic concern is the potential impact of climate change on Australia's agricultural sector. Other impacts predicted would adversely affect Australia's coastal zone, its fragile ecosystems and its biodiversity. There are many more. As Penny Wensley, current Australian Ambassador for the Environment, has said, 'It is this situation of double jeopardy – of action or inaction on climate change – that dictates the logic of our [Australia's] approach to the negotiations overall' (Department of Foreign Affairs and Trade 1992b).

The Australian government thus has dual goals to achieve in climate change policymaking. Bolin has acknowledged the overriding nature of domestic interests in greenhouse policy negotiation at the international level:

It is essential in the case of an environmental issue, such as global climate change, that politicians consider the global aspects of the problem, and at the same time serve as representatives of their respective governments. They are faced with the task of how to protect the interests of their own countries as well as possible. Generally, this is an overriding instruction to any delegate that takes part in international negotiations .... In the long-term, the IPCC assessments should aim to convince politically responsible national bodies to consider the global and long-term aspects of the issue and to formulate instruction to delegates accordingly.

(Bolin 1994, pp. 27–8)

Notwithstanding Australia's concerns about protecting its own domestic energy industries, Ros Kelly urged other nations to take up the greenhouse policy challenge at an IPCC workshop in Canberra in 1992 sponsored by the Australian government; she said:

we cannot wait for the research to provide us with a comprehensive, certain picture. The wisest choice is surely to adopt the precautionary principle and to instigate response measures in the knowledge that certain consequences are likely to flow from climate change.

Certainly, there are few excuses not to begin immediately to implement the so-called 'no regrets' measures as a major part of a first-generation greenhouse response strategy.

(IPCC 1992a, p. 15)

Also, at the United Nations Conference on Environment and Development (UNCED) in Rio in June 1992 in Australia's address to the Conference, Ros Kelly called for all nations to ratify the climate change and biodiversity conventions as soon as possible. She said, 'much more needs to be done than just signing the two conventions at the summit as neither is an end result in itself' (Weekend Australian 1992).

Australia's record on the implementation of greenhouse policy at the domestic level and the constraints faced on implementation are discussed further below.

TECHNOCRATIC POLITICS, PUBLIC OPINION AND GREENHOUSE

Australian public opinion on the environment and greenhouse effect

Two recent papers focusing on Australian public opinion on environmental issues are those of McAllister and Studlar (1993) and Pakulski and Crook (1995). McAllister and Studlar (1993) comment that 'In line with trends in public opinion in other countries, environmental issues in Australia rose to prominence in the late 1980s.' They assert that

> This increased popular concern with the environment was due, in part, to international factors such as the Chernobyl nuclear accident and the dissemination of information about global warming. But there were also domestic Australian environmental issues ... which ... sensitized public opinion to these issues.
>
> (McAllister and Studler 1993, p. 354)

Pakulski and Crook likewise have observed that

> Between 1988 and 1990 the proportion of people for whom the environment was the issue of most concern jumped from 5 to 26 percent.... This jump in the popular level of concern – which persisted through 1990 and 1991 – amounted to a revolution in popular environmental consciousness. The major boost for this revolution came from enormous publicity given to environmental issues, especially the 'greenhouse effect', by the media, as well as from political change. Green issues made a gradual entry into the Party agenda .... In 1993 green concerns seem to have subsided [21 percent saw environment as the issue of most concern]. However, careful scrutiny reveals a persistence of environmental concerns which is remarkable given the low level of publicity during the [1993] election campaign.
>
> (Pakulski and Crook 1995, p. 40)

Pakulski and Crook highlight the significance of the time-scale of environmental change versus the political time-scale. They say from the results of the 1993 Australian Electoral Study survey of 1,780 respondents (a representative sample of the Australian adult (voting) population), collected immediately following the 1993 federal election, that

> the concerns persist outside the immediate party/election agenda. When asked to nominate the issue of most concern in the longer time perspective ('10 years from now') 22 percent of respondents nominated the environment as the first or second most important concern thus propelling it to a close third place behind 'unemployment' and 'health' ... [which], unlike environmental issues, were at the top of party-

election agenda and therefore received massive publicity in the mass media.

(Pakulski and Crook 1995, p. 41)

Pakulski and Crook (1995) conclude that 'concerns about the environment are located by the [Australian] public in a long time-frame' and that 'environmental concerns have entered the public consciousness and are likely to stay there'. Their analysis also shows that environmental concerns have spread through broad sections of the population 'outside the initial circles of young and educated city dwellers', and they say that it gives some indication that Australians 'are all green today'.

The degree of concern expressed by Australians about the greenhouse effect is very interesting: 71 per cent of respondents in 1990 and 66 per cent of respondents in 1993 saw the greenhouse effect as a 'very urgent' environmental issue while 19 per cent in 1990 and 17 per cent in 1993 saw it as the 'most urgent' environmental issue (Pakulski and Crook 1995). The greenhouse effect ranks second only to 'pollution' for Australians in terms of perception of degree of urgency of the issue.

The greenhouse effect, public opinion and the media

As Pakulski and Crook (1995) indicate, the media in Australia played a major role in raising public consciousness about the greenhouse effect in the late 1980s. More recently, there has seen some concern about the role of the media in terms of polarising the arguments of climate change experts into pro- or anti-greenhouse camps and thus reducing public concern about climate change (see, for example, Henderson-Sellers 1993). This phenomenon is not exclusive to Australia; for example, Schneider (1991), in reference to the situation in the United States, has expressed the concern that 'a highly confusing and polarising media debate can be paralysing to anticipatory management' for the greenhouse effect. Also, Bolin has said: 'Interpretation must not be left, exclusively, to the journalists, because matters tend to become exaggerated and, if repeated frequently, may mislead the public' (van Zijst 1994, p. 5). Interestingly, the percentage of Australians concerned about the greenhouse effect appears not to have diminished significantly even given their exposure via the media to greenhouse 'doubters'.

The impact of public opinion in greenhouse policymaking

It is also interesting to compare these Australian public opinion findings with Rhodes's (1992) pessimism about the impact of public opinion in the United States on greenhouse policymaking in the long term. He says:

> Preserving global and regional climate resources or seeking to control the rate or scope of change constitute almost unimaginable aspirations

129

... in the long term it will be difficult to sustain popular aspirations on such a large scale to preserve an asset such as climate that is difficult to define, and one which is so difficult to assess at any given moment as to its stability or variability. That is, how might one demonstrate that progress towards climate change stabilization or postponement was being achieved and over what time-frame?

(Rhodes 1992, pp. 210–11)

While Crook and Pakulski's findings about the potential for a persistence of public concern about climate change, in the Australian context, are encouraging, assuming that sustaining the public interest is a necessary major catalyst for policy action on the greenhouse effect is debatable. Public interest is undoubtedly a factor influencing climate change policymaking, but a public-interest-driven model of policymaking (for example, Downs's (1972) 'issue-attention cycle') is an over-simplification of the complex driving forces on government action on an issue such as the greenhouse effect that is based on scientific information. As Beer contended in reference to such scientifically based issues, government action in the United States and other Western nations is subject to 'the growing complexity of knowledge which is used in identifying and solving contemporary problems and ... an ever-greater dependence on experts in policymaking' (Beer 1973, p. 75).

Climate change politics as technocratic politics

Beer (1973) argues that elected governments in the domain of technocratic politics[8] are subject more to the influence of their scientific and bureaucratic advisers and clientele groups than to public interest groups and to the electorate as a whole. Beer does caution, however, that the base of support for technocratic policies within the electorate and interest groups is not completely insignificant. Technocratic politics thus principally assumes an élitist pattern of political behaviour. Schneider's observations about greenhouse policymaking have congruence with the technocratic politics model. He says:

If the public is totally ignorant of the nature, use, or validity of climatic (or many other kinds of) models then public policy-making based on model results will be haphazard at best. In this case the decisionmaking process tends to be dominated by special interests or a technically trained élite.

(Schneider 1991, p. 52)

AUSTRALIA'S DOMESTIC POLICY RESPONSE

The National Greenhouse Response Strategy

In Australia the political will to act on the greenhouse effect has been influenced significantly by the imperative of delivering policy that is politically acceptable in the short term. Australia's *National Greenhouse Response Strategy* (Australian Government 1992) was released in December 1992. The strategy was produced via a two-stage process over two years. The first phase of the policy development process for climate change in Australia involved the parallel production of, first, the Ecologically Sustainable Development (ESD) *Greenhouse Report* (ESD 1992) by an Australian government-sponsored intersectoral ESD Greenhouse Coordinating Group and, second, a report by the Industry Commission, titled *Costs and Benefits of Reducing Greenhouse Gas Emissions* (Industry Commission 1991). Both the ESD Working Groups and the Industry Commission were asked by the Australian government in 1990 to report on costs, benefits, and options for reducing greenhouse gases. However, their individual reporting mandates were somewhat different.

More specifically, the then Prime Minister, Bob Hawke, asked that 'in the context of finalising the government's ESD Strategy, the working groups report on options to achieve the interim targets and the most cost effective combination of measures needed to achieve these targets' (ESD 1992). The ESD report was produced via a consensual 'round table' process involving input from the bureaucracy (state and federal), industry, trade unions, the scientific community, environment groups and other community interest groups (ESD 1992).

Differing in orientation from the ESD *Greenhouse Report*, the Industry Commission report was derived from an inquiry process involving input from interested parties through submissions and public hearings. The terms of reference of the inquiry required the Commission to report on:

(a) the costs and benefits for Australian industry of an international consensus in favour of a stabilisation of emissions of greenhouse gases not controlled by the Montreal Protocol on Ozone Depleting Substances, based on 1988 levels, by the year 2000 and a reduction in those emissions by 20% by the year 2005;

(b) the opportunities that could arise for Australian industry as a result of that international consensus; and

(c) how Australia would best prepare itself to respond to those costs and benefits.

(Industry Commission 1991)

The findings of the Industry Commission were well received by industry – particularly the energy and mining-related sectors and, in particular, the coal sector (Australian Coal Industry Council 1992) – and unanimously rejected by environment groups (see, for example, Diesendorf 1992). The report needs to be considered in the context of the Commission's charter. As Industry Commissioner Tor Hundloe emphasises:

the Commission can only go so far in analysing and making recommendations on environmental/sustainability issues. It, like any other such constituted body, has to work within its statutory charter; in its case, particularly the policy guidelines which emphasise economic efficiency.

(Hundloe 1992, p. 476)

According to Hundloe (1992), 'In attempting to meet its terms of reference – compare costs and benefits – the Commission was forced to state there was no technical solution.' He also said that the intergenerational equity considerations associated with the greenhouse issue were too difficult to deal with, being beyond the scope of economics, and that the limits of the discipline need to be recognised.

In the Industry Commission report, the uncertainties in climate change science are emphasised and used to argue that Australia should not act on greenhouse warming. Hundloe's assessment of the Commission's difficulty with its brief was that 'the economists cannot measure impacts if scientists cannot tell them what the impacts are likely to be' (Hundloe 1992). The Commission argued that as other nations are unlikely to act cooperatively on the basis of an international consensus regarding the need to act on climate change, unilateral action by Australia is not in the national interest. Additionally, it is reasoned that there would be costs to Australia in participating in regionally limited bilateral or multilateral agreements. The Commission reduced national output by 1 per cent and concluded that most sectors of the economy would be adversely affected by emissions reduction and that the burden of the costs and adjustments in particular industries (notably coal) and regions would be significant.

By contrast, the comprehensive ESD *Greenhouse Report* provides an extensive list of greenhouse-relevant policy recommendations in the areas of energy production, energy use, transport, manufacturing, mining, agriculture, fisheries, and tourism. In particular, the recommendations made in the areas of energy use and transport are proposals that ought to have a significant impact on reduction of greenhouse gas emissions. But the ESD Forest Use Working Group did not make any recommendations specific to climate change. In addition to the sectoral recommendations, the National Greenhouse Steering Committee recommended in the *Greenhouse Report*

that ... the detailed greenhouse-related recommendations from the ESD Working Groups ... form the basis of a National Greenhouse Response Strategy.

... that in conjunction with the Commonwealth/State ESD Committee, the National Greenhouse Steering Committee draw up clearly defined responsibilities and an agreed time-table for implementation of these recommendations at Commonwealth, State, Territory and local government levels, for approval and adoption by all governments by no later than mid-1992.

(ESD 1992, p. xxvi)

The second stage of the greenhouse policy development process in Australia has been referred to by Mark Diesendorf of the Australian Conservation Foundation (one of two environment group representatives on the ESD Greenhouse Coordinating Group) as 'the bureaucrats' betrayal' (Diesendorf 1992). First, a committee of state and federal bureaucrats, the National Greenhouse Steering Committee (NGSC), watered down the recommendations of the ESD *Greenhouse Report* (ESD 1992) to produce a Draft National Greenhouse Response Strategy (NGSD 1991). Diesendorf has said: 'during the first phase of the ESD process, some of these bureaucrats had already been attempting to undermine the moves by industry, trade union and environment movement representatives ... in Phase 2 of the ESD process, the bureaucrats had a free hand' (Diesendorf 1992). The National Greenhouse Steering Committee was made up of officials representing all tiers of government in the federal system down to the local government level and had three subcommittees reporting to it. Some lip-service consultation was made with ESD representatives with regard to the draft strategy after its circulation for public input but little response was made to criticisms and a further weakened final policy document was finally released: the *National Greenhouse Response Strategy*.

The Strategy is couched in the terms of 'no regrets' actions which, as mentioned at the beginning of this chapter, Ros Kelly has urged other nations to take; but, added to this philosophy, has been the core objective that 'equity considerations should be addressed by ensuring that response measures meet the broad needs of the whole community and that any undue burden of adjustment potentially borne by a particular sector or region is recognised and accounted for' (Australian Government 1992, p. 12). This guiding principle indicates that any industry sector (for example, the mining, chemical, or petroleum sectors – all large greenhouse gas producers) or any single geographic region should not be economically burdened. This differs from the ESD *Greenhouse Report*, which reasoned that some industry decline and closure and some restructuring would be necessary to achieve cuts in greenhouse gas emissions but argued that this would be countered by growth in industry oriented towards energy efficiency and renewable energy.

The Strategy has thus evolved into a plan of least possible change. It has disappointed environmental groups and has been said to be evidence that 'Australia is becoming more and more part of the fossil fuel club'

(Greenpeace Australia 1992). Overall, the Strategy concentrates very much on reviewing initiatives of the various tiers of government already under way and confirming that these will continue. It commits state and territory governments only to relatively small and fragmented new initiatives. Accordingly, it is a cautious document and maintains the status quo. State and territory governments are not required to identify any definite time frames or targets to reduce greenhouse gas emissions. This is notwithstanding that Australia's Interim Planning Target is described as 'a yardstick against which the implementation of greenhouse measures can be assessed' and as 'a focus for action in Australia to help mitigate human caused global climate change and associated adverse impacts' (Australian Government 1992, p. 8).

The Strategy document includes an estimate of the extent of emission reductions that may result from implementation of its recommendations. With respect to CO_2, reductions in emissions from energy end-use sectors are estimated to be in the range of 10 to 95 Mt per year by 2005 (Australian Government 1992), with the comment being made that 'It seems likely ... that the reduction achieved would be towards the lower end of the range' (Australian Government 1992, p. 92). Also, as a developed country party to the UN Framework Convention on Climate Change, Australia is expected to aim to reduce greenhouse gas emissions to 1990 levels by the year 2000 and to report on progress towards such stabilisation in the meantime. To achieve the Convention aim, the high end of the range would need to be achieved at least. The Strategy also indicates that if the Interim Planning Target is to be achieved, emissions of CO_2 would need to be lower by 200 Mt/yr by 2005 (Australian Government 1992). This assumes equivalent reductions in emissions for all greenhouse gases. In reality, cuts for CO_2 should be higher as it is relatively easier to implement cuts in CO_2 emissions than for other greenhouse gases.

Factors shaping Australia's domestic response

There are many obstacles to overcome with respect to Australian domestic greenhouse policy if Australia is to fulfil its obligation as a signatory to the Climate Convention and this is notwithstanding the fact that the Convention is an evolutionary document. The apparent barriers to achieving this stabilisation goal for Australia include the following.

Scientific, economic and policy uncertainties

Considerable coordination of scientific research regarding the nature, scope, and severity of impacts of global climate change has been carried out since the inception of the IPCC. However, scientific uncertainty remains about the nature of climate change and its impacts, and while uncertainty exists the political impetus to implement effective greenhouse policy measures will be hampered. Also, from the political decision-maker's perspective, uncertainty exists with respect to the costs and benefits of greenhouse policy options both to mitigate climate change and for adaptation to it. It is not surprising then, given the long-term policy horizon of decades and with the probability of major costs only being realised in the future, that there is a lack of political will to take effective action now on the greenhouse effect.

Australian economic dependence on the energy sector

The potential implications of reducing greenhouse gas emissions in Australia given Australia's economic dependence on the energy sector are also a concern for political decision makers. Significant cuts in CO_2 emissions, which would be necessary to achieve the Convention's aim to reduce emissions to 1990 levels by the year 2000, would necessarily involve more than the 'no regrets' reductions of the National Greenhouse Response Strategy in the use of the CO_2-intensive fuels: coal and oil. According to Lowe (1989), state electricity authorities (running coal-fired power stations) and oil companies together are responsible for about 80 per cent of Australian CO_2 emissions. The concern is that Australia could lose significantly in the short term and, as mentioned previously, the caveat to the Interim Planning Target indicates that Australia is unwilling to shoulder such costs if other developed nations take little action. The federal government has faced concerted lobbying from groups such as the New South Wales Coal Association, the Australian Mining Industry Council, and the Business Council of Australia on this issue (McKanna 1992). The influence of these groups conforms with Beer's (1973) description of the role of clientele groups in technocratic politics. Environmental groups in Australia, and in particular the Australian Conservation Foundation and Greenpeace, have lobbied intensively on the greenhouse issue but have been less successful.

Federalism and bureaucratic inertia

The division of federal–state powers in Australia has been a major stumbling-block in terms of formulating an effective national climate change strategy. As environmental protection is under the shared jurisdiction of the states and the federal government, with the latter having basically a watchdog capacity, the environmental and economic orientations of the states played a large part in the shaping of the *National Greenhouse Response Strategy*.[9] New South Wales's representatives, for example, reportedly played a particularly obstructive role in terms of developing the strategy, and Western Australia refused to report on the first year of implementation of the strategy in December 1993 (NGSD 1994). Bureaucratic obstruction and inertia have also proved to be a problem in implementing greenhouse policy. The Australian

Department of Primary Industry and Energy was charged in 1993 with poor performance in implementing the preliminary greenhouse policy undertakings of the Australian government announced in 1990, despite the sense of priority intended by the government (Bowden *et al.* 1993); this is illustrative of problems that can occur in federal agencies and equally, if not more importantly, at the state level. It is not an easy task for government agencies that have been oriented towards facilitating energy production and consumption to reorient to implementing energy efficiency measures and conservation. Their long-standing relationship with energy industry clientele groups compounds this problem.

These factors indicate that the climate change policymaking process is proving to be extremely challenging at the domestic level in Australia. Significant political and economic barriers exist. Accordingly, the current national greenhouse policy is very much a 'wait and see' approach and does not directly address Australia's high per capita emissions levels and does not give an impetus for change.

CONCLUDING DISCUSSION

Time-related factors of influence in the climate change policy process include the following: scientific progress made with global climatic modelling; election cycles and domestic political events; international political commitments to time-frames for action, for example, the Toronto Target or the Climate Convention 'aim'; long-term strategies of investment in technology (for example, electricity generation); economic development and investment cycles; and changes in the law (for example, the potential production of a Climate Convention Protocol with targets and timetables for the reduction of CO_2 emissions). These factors all have the potential to sway political will both internationally and domestically to either strengthen or weaken greenhouse policy.

To date, Australia has played a reasonably proactive role in the climate change policy arena internationally. Not only in the climate change policy area, but in many aspects of environmental concern, Australia has arguably played the role of good environmental citizen. Australia has, in general, gained the respect of its developing neighbours as well as other developed nations with regard to its handling of environmental questions. However, the political will in Australia to act at the domestic level on greenhouse warming has been influenced significantly by the imperative of delivering policy that is acceptable in the short term. Accordingly, success to date in developing effective policy to address climate change has been limited. This problem is not unique to Australia; the climate change policymaking process is proving to be extremely challenging at the domestic level in many countries owing to the political, economic, and time-scale complexities of the climate change issue.

In the words of the Convention: 'the global nature of climate change calls for the widest possible cooperation by all countries and their participation in an effective and appropriate international response'. The time-scale of this global response is dependent on the domestic response of nations. According to IPCC predictions, domestic policy responses to the greenhouse effect need to be implemented now to reduce greenhouse gas emissions in order to avert the worst consequences of climate change.

Examination of greenhouse policy formulation and implementation in Australia to date indicates that Australia's domestic policy focus is on short-term economic impacts on particular industry sectors or geographic regions; this short-term perspective has prevented the development and implementation of climate change policy that focuses on the broader global consequences or the longer-term time-scale of greenhouse warming. Ironically, survey results indicate that two-thirds of Australian voters see the greenhouse effect as a very urgent environmental issue. This situation of significant public concern and limited policy response reflects the technocratic nature of greenhouse policymaking. Approaches to environmental policymaking such as ecologically sustainable development, the precautionary principle and intergenerational equity, do prove conceptual tools that address the time dimensions of the climate change issue and its uncertainties. They have been used symbolically in the development of Australian greenhouse policy, but they have had minimal impact on the final shape of the policy because of the domestic politics of the issue.

Professor Charles Birch, an emeritus professor of Sydney University, has said: 'it is impossible for Australia, and for that matter other countries, to survive and flourish except as they find their future in a global perspective' (Birch 1993). Addressing climate change is proving to be a testing ground for whether a long-term perspective of human impact on the global environment can be incorporated into domestic policy.

NOTES

1   The major long-lived greenhouse gases in the atmosphere are carbon dioxide (CO_2), methane (CH_4), nitrous oxide (N_2O), chlorofluorocarbons (CFCs), and carbon tetrachloride (CCl_4).
2   This doubling of CO_2 concentration was anticipated to occur by 2030 at the time of the Villach conference in 1985.
3   The Convention as a whole is a modest document and while being an important step in international recognition of climate change as a problem, it contains no concrete commitments on stabilisation and reduction of emissions of carbon dioxide and other greenhouse gases. The text of the Convention lacks teeth primarily because of the United States' non-negotiable stance on commitments.
4   Climate change predictions at the regional scale are not yet available, the current grid scale of global climatic models being of the order of 500 × 500km.
5   International initiatives which Australia has been involved in include protection

of the earth's biodiversity, reduction of ozone depletion, environmental protection of Antarctica, prevention of international trade in endangered species of wild fauna and flora, protection of World Heritage, responding to global climate change, and many others. Australia also gives environmental management and training assistance to developing nations through the Australian International Development Assistance Bureau (AIDAB) and has established bilateral environment programmes with South Pacific nations, Association of South East Asian Nations (ASEAN) countries and Papua New Guinea (see Department of Foreign Affairs and Trade 1992a).

6 Agreed at the 1988 Conference on the Changing Atmosphere held in Toronto, Canada.

7 Australia ratified the Framework Convention on Climate Change on 31 December 1992. The Convention came into force on 26 March 1994, 90 days after fifty nations had ratified it.

8 Beer (1973) uses the term 'technocratic politics' to refer to the process of policymaking that incorporates scientific information, and describes it as operating mainly within what he calls a professional–bureaucratic complex. Four major groups function in this complex. These are the research élite, the professional sectors of the bureaucracy, clientele groups, and elected governments. Members of the first of these groups, the research élite, reside in universities, research institutions, and consultancies, and have close connections with the professional sectors of the bureaucracy; they have considerable influence. The professional sectors of the bureaucracy are the second major sphere of influence. They embrace concepts put forward by the research élite and make suggestions to the research élite about new areas for applied research. They also liaise with clientele groups, the third important sphere of influence. Differing from traditional interest groups, clientele groups rely on technocratic government programmes. A final major sphere of influence is composed of elected governments at all relevant levels. They act on the advice of the scientists and professionals, liaise with clientele groups and other interests groups, and take into consideration the mood of the electorate and assist in mobilising electoral consent.

9 This is notwithstanding the existence of the Intergovernmental Agreement on the Environment (IGAE), which, although signed in May 1992, in practice has not been implemented. Environmental lawyer Professor Rob Fowler's opinion is that it is doubtful that it will be implemented and that the fundamental problem with the IGAE is that achievement of change through an informal document with no legal base is difficult (Fowler 1993).

REFERENCES

Australian Coal Industry Council. 1992. *Responses to the Greenhouse Effect: A Coal Industry Perspective*. Canberra: ACIC.

Australian Government. 1992. *National Greenhouse Response Strategy*. Canberra: Australian Government Publishing Service.

Barnett, D. 1990. Ros Kelly's mission – the getting of greendom. *Bulletin*, 30 October, 32–4.

Beer, S. 1973. The modernization of American federalism. *Publius* 3(2): 74–9.

Birch, C. 1993. *Confronting the Future – Australia and the World: The Next Hundred Years*. Ringwood: Penguin.

Boardman, R. 1991. Approaching regimes: Australia, Canada and Environmental Policy. *Australian Journal of Political Science* 26(3): 446–71.

Bolin B. 1994. Science and policy making. *Ambio* 23(1): 2–29.
Bowden, J., Fitzerale, R., Chapman, A., and Prentice, M. 1993. *Efficiency Audit: Implementation of an Interim Greenhouse Response.* Department of Primary Industries and Energy Energy Management Programs, Australian National Audit Office, The Auditor-General Audit Report No. 32, 1992–3. Australian Government Publishing Service.
Christie, I. 1992. Social and political aspects of global warming. *Futures* 241: 83–90.
Department of Foreign Affairs and Trade, Australia. 1992a. Fact-Sheet on Australia: Australia and the Environment. Overseas Information Branch, Canberra, March.
Department of Foreign Affairs and Trade, Australia. 1992b. Australia: a lot to lose from climate change. *Environment: Australia's International Agenda* 4 (April): 3.
Diesendorf, M. 1992. 'Integrated Greenhouse Policies for Energy and Transport', Institution of Engineers Australia's Greenhouse Policy Seminar: *Proceedings*, Canberra, 26 June.
Downs, A. 1972. Up and down with ecology – the 'Issue-Attention Cycle'. *Public Interest* 28: 38–50.
Edelman, M. (1964) *The Symbolic Uses of Politics.* Urbana: University of Illinois Press.
ESD (Ecologically Sustainable Development) 1992. *Greenhouse Report.* Canberra: Australian Government Publishing Service.
Evans, G. and Grant, B. 1991. *Australia's Foreign Relations in the World of the 1990's.* Carlton: Melbourne University Press.
Fowler, R. 1993. 'New National Directions in Environmental Protection and Conservation'. Australian Centre for Environmental Law Environmental Outlook Conference, Sydney, 10–11 November.
Godard, O. 1992. Social decision making in the context of scientific discoveries – the interplay of environmental issues, technological conventions and economic states. *Global Environmental Change* (September): 239–49.
Greenpeace Australia. 1992. Keating's recipe for escalating greenhouse pollution. *Media Release*, 1 July.
Henderson, J. 1992. Perspectives on UNCED. *Climate Change Newsletter* 4(3): 2.
Henderson-Sellers, A. 1993. Director's viewpoint. *Climatic Impacts Centre, Annual Report 1993*, Macquarie University, Sydney.
Hundloe, T. 1992. The role of the Industry Commission in relation to the environment and sustainable development. *Australian Journal and Public Administration*, 51(4): 476–89.
Industry Commission, Australia. 1991. *Costs and Benefits of Reducing Greenhouse Gas Emissions*, vol. 1: Report, Industry Commission Report No. 15, Canberra: Australian Government Publishing Service.
IPPC [Intergovernmental Panel on Climate Change]. 1992a. *Climate Change: IPCC Response Strategies Working Group – Proceedings of a Workshop on Assessing Technologies and Management Systems for Agriculture and Forestry in Relation to Global Climate Change.* Canberra: Australian Government Publishing Service.
IPCC [Intergovernmental Panel on Climate Change]. 1992b. *Climate Change 1992: The Supplementary Report to the IPCC Scientific Assessment.* Cambridge: Cambridge University Press.
Jager, J., and Ferguson, H. 1991. *Climate Change: Science, Impacts and Policy – Proceedings of the Second World Climate Conference.* Cambridge: Cambridge University Press.
Lowe, I. 1989. *Living in the Greenhouse.* Newham: Scribe.
McAllister, I., and Studlar, D. 1993. Trends in public opinion on the environment in Australia. *International Journal of Public Opinion Research* 5(4): 353–61.

McKanna, G. 1992. Greenhouse dilemma. *Airways*, May/June, pp. 75–6.

McPhail, I. 1992. Rio – a Commonwealth perspective. *Environment Institute of Australia Newsletter*, Special Issue 'Report from Rio', September, pp. 14–16.

NGSD (National Greenhouse Steering Committee). 1991. *Draft National Greenhouse Response Strategy*. Canberra: NGSD.

NGSD (National Greenhouse Steering Committee). 1994. *Summary Report on the Implementation of the National Greenhouse Response Strategy*. Canberra: Commonwealth of Australia.

Pakulski, J. and Crook, S. 1995. Shades of Green: Public Opinion on Environmental Issues in Australia. *Australian Journal of Political Science* 30: 39–55.

Rhodes, S. 1992. Climate change management strategies: lessons from a theory of large scale policy. *Global Environmental Change* 2 (September): 205–14.

Richardson, G. 1990. Greenhouse – the challenge of change. In Coghill, K. (ed.) *Greenhouse: What's to Be Done*. Leichhardt: Pluto Press.

Schneider, S. H. 1991. Prediction of future climate change. In Tester, J., Wooe, D., and Ferrari, N. (eds) *Energy and the Environment in the 21st Century*. Cambridge, MA: MIT Press.

Schulman, P. 1980. *Large-Scale Policymaking*. News York: Elsevier North-Holland.

Taplin, R. 1994. International policy development on greenhouse and the involvement of the island South Pacific. *Pacific Review* 7(3): 271–82. (Special issue, on 'The UN and the Pacific'.)

van Zijst, P. 1994. 'Politicians deal with politics, scientists better not get mixed up with that': Interview with Professor Dr Bert Bolin. *Change* 18 (February): 4–7.

Weekend Australian. 1992. Week in review: national news. *Weekend Australian* 13–14 June, p. 25.

Zillman, J. 1990. Second World Climate Conference. *Climate Change Newsletter* 2(4): 2–3.

7

CONFLICTING
TIME-SCALES
Politics, the media, and the environment
John Gordon

Editors' note In this chapter, in one sense John Gordon is optimistic: he believes that there is an increasing realisation by both government and industry that there has to be some kind of longer-term planning. On the other hand he is pessimistic, believing that in the democracies, the political necessity of short-term thinking dominates most other requirements. This is not the same for all cultures and all democracies, some being better able than others to think long term – but in Britain the political system is seen as being particularly unhelpful, given the adversarial nature of the two-party system. The non-governmental organisations (NGOs) are not much help in this situation either: their usual tactics involve very current issues and very current demands – and had they been able to think more long term and strategically, they could have had a much greater impact. But they too are caught by an uncertain future, which requires them to act while they can.

INTRODUCTION

My theme is the linkage between time-scale and the perceptions, formulation, and implementation of public policy as seen through the eyes of someone who spent many years in government. My perspective is Northern, and largely British.

Concern about the implications of time-scale for government is of course nothing new. In the early 1960s President Kennedy likened the Washington system to a rather stupid and ill-coordinated dachshund, 'whose head hung down with tears and sadness' while the message got stuck somewhere in the middle and 'the tail wagged on in previous gladness'.

This chapter looks at the time-scale of current conventional politics, to suggest some reasons for it and also some exceptions. It analyses some of the factors which are compelling a longer time-scale, above all that of sustainable development. It examines some of the key factors influencing this change.

Finally it suggests some ways in which the process of political change in British politics needs to incorporate the challenge of longer time-scales.

THE LIMITATIONS OF THE ELECTORAL CYCLE

At one level the problem is simple. Politics in Britain and most other democratic countries is overwhelmingly centred round short-term issues and two- to five-year electoral cycles. Politicians need to be re-elected to survive, and by the very nature of the traditional political process everything is subordinated to this interest. Almost inevitably, therefore, only sporadic attention is paid to longer-term issues which can in political terms be safely postponed without immediately dangerous practical consequences. Thus, for example, short-term issues of recession, unemployment, and crime at home and regional conflict abroad inevitably take priority over addressing such long-term issues as global pollution, resource depletion, and population increase. This perspective is of course also reflected in the workings of finance and industry, above all in the 'Anglo-Saxon model' in operation in Britain and the United States. Rather longer-term time-scales prevail in the more enlightened capitalism of Germany and Japan.

Yet even in traditional politics some important decisions have to be made against a far longer time-scale. These relate largely to patterns of infrastructure development where for practical reasons politicians and planners need to operate far beyond the usual electoral cycles. Local structural plans in Britain governing land use patterns are designed to last for fifteen to twenty years. Road building and other major projects, notably power stations, also require a similar time-scale from planning through to commissioning. Such processes are inevitably based on a number of implicit or explicit assumptions about the nature of economic development.

In recent years, however, these assumptions have been increasingly challenged by the environmental movement in terms, for example, of land use and transport policies which promote the continuing rapid disappearance of the English countryside or long-term energy policies which ignore the need for energy conservation. Thus, for example, public inquiries on individual projects have on occasion turned into major confrontations on conflicting long-term perspectives. The most notable example in recent years has been the inquiry on the proposed Sizewell B nuclear power station. Conversely such confrontation very rarely, if at all, happens in inquiries on new roads because the Department of Transport's terms of reference explicitly rule out raising wider questions of transport policy (leading some environmental campaigners to liken the DoT position to one of enquiring whether the public preferred to eat babies boiled or fried without allowing them to say that whether or not they wanted to eat babies in the first place, meaning that the public is consulted on the alignment of a by-pass but not on the need for it in the first place).

In parallel, a growing challenge to short-term time-scales has been mounted internationally as Britain and other countries have been increasingly involved in negotiating long-term environmental agreements of one sort or another. Britain's first formal commitment into the twenty-first century, on limiting national sulphur emissions from power stations, was undertaken under the 1983 Economic Commission for Europe Convention on Long Range Transboundary Air Pollution. Similar long-term commitments on conservation of habitat and species were negotiated under the United Nations Environment Programme (UNEP). This revised approach to time-scales was greatly reinforced at the 1992 Rio Earth Summit, when 108 heads of state and government entered into a partnership 'reflecting global consensus and political commitment at the highest level' to pursue sustainable development into the next century. Agenda 21, the main end product, contains a multiplicity of individual long-term targets. These include no fewer than fifty-eight to be achieved in or by the year 2000. Rio's two other major end products, the Framework Convention on Climate Change and the Convention on Biological Diversity, are not specifically linked to time-scales but by their very nature, and that of the problems they are addressing, are also clearly seen as very long-term commitments.

TIME-SCALE REVOLUTION

Are we now, under the influence of Rio, now entering a new stage of politics, when longer-term considerations begin to be reflected more centrally? If so, the key concept is that of sustainable development. By definition it is time based. In the classic Brundtland Report definition it is seen as 'development which meets the need of the present without compromising the ability of future generations to meet their needs'. The concept has been happily accepted by all the governments of the world. In theory at least, intergenerational equity is therefore part of the political acquis of all societies.

Closely connected has been the rediscovery of the importance of long-term planning. This is a central, if implicit, theme of Agenda 21. The UK national strategy for sustainable development and matching strategies on climate change and biodiversity published by the British government in January 1994 carefully avoid the term and fall short of specific, measurable commitments. But they take a twenty-year perspective and mark an important stage in the government's conversion to the view that relevant policies must be pursued over a far longer-term time-scale than in the past. In parallel, the European Union's 5th Environmental Action Programme sets out a coherent framework of long-term policy objectives and measures designed to set the Union on a transitional path to sustainability by the year 2000.

But all this remains highly theoretical. Listening to politicians or reading the papers you would be very hard pressed to notice any difference. Vested interests opposed to change remain immensely powerful. There is very little sign, for example, that most economists, let alone bankers or business people, have accepted the message. Most indeed have not yet heard it. Nor indeed probably have more than a small minority of the electorate. Fewer still are yet prepared to act on it. The overwhelming temptation for government is to treat this new approach to time-scales in terms of what psychologists would call a displacement activity, to consider the national strategy as a sort of household god which can be whisked out occasionally for ceremonial invocation and ignored the rest of the time.

PROSPECTS FOR FURTHER CHANGE

Against this background, what are the forces promoting longer time-scales in national politics? Mainstream politicians are often very open, indeed self-critical, in private about the short-term nature of their preoccupations. But they are too dominated by electoral cycles to be able themselves to lead the way. Think-tanks of both right and left have largely ignored the problem, as have pressure groups promoting political and constitutional reform. Can we look to the environmental pressure groups, large and powerful organisations with a bigger membership than that of the main political parties, to press effectively for a more enlightened approach? And to what extent are the media making people more aware of the inadequacies of current political time-scales?

There is no need here to sing the praises of the environmental movement. Its achievements in terms of putting environmental issues on the agenda and forcing governments to take action have been immense. But sadly the relentlessness with which green pressure groups shadow government policy often comes to reflect the same time-scales. Thus they become the victims of the short-termism which they so fiercely and rightly criticise in government. Their own planning processes still rarely go beyond two or three years. All too often they bring pressure to bear at the closing stages of the process they are trying to influence, once the key decisions have been taken. Very often the scope is far greater earlier, when options are more open and government's investment in one particular outcome is less engaged.

One example of this 'failure of time-scales' is the unsuccessful campaign led by Greenpeace and FoE against the Sellafield Thermal Oxide Reprocessing Plant (THORP) in 1993. The go-ahead for THORP was given after a public inquiry and parliamentary debate in 1977. The major NGOs most concerned, however, lost interest afterwards and refocused on THORP only when it was almost completed, some sixteen years later. By then so much money and political capital had been invested in it that any decision not to go ahead would inevitably have been seen as a major defeat for the government. By contrast, had Greenpeace and FoE struck again in, say, the middle 1980s when the project was bugged by cost overruns and there were major doubts in Whitehall about the wisdom of going ahead as uranium prices fell and the economic rationale weakened, the chances of success would almost certainly have been far greater.

A second example of pressure groups' neglect of time-scales can be seen in their handling of the General Agreement on Tariffs and Trade (GATT) Uruguay Round. Environmental and developmental NGOs began seriously to express their alarm at the GATT Uruguay Round's failure to take environmental issues into account only in the early 1990s, when negotiations were almost completed and almost all governments were under very strong pressure to finalise them without further delay. A coalition of pressure groups mobilised to alert public and Parliament to the dangers and succeeded in forcing a debate in the House of Commons. But the ground had not been prepared sufficiently in advance and the outcome was never seriously in doubt. If pressure groups had acted half as decisively when the round was beginning in the early 1980s they would perhaps still not have been able to prevent its being completed. But they would almost certainly have been able to insist on environmental criteria being incorporated in the negotiation process and thus ensured an outcome less damaging to the political credibility of the NGOs concerned and perhaps also to the global environment.[1]

The media, above all the 'heavyweight' press and television, have a far more central role than the environment groups in defining the national political agenda. Do editors, producers, and journalists think further ahead, or at least criticise politicians for almost exclusive preoccupation with the short term? There is unfortunately very little, if any, evidence that they do. Of course there are some excellent newspaper articles and TV programmes warning us of the long-term problems ahead. But these are only occasional, and drowned in an ocean of preoccupation with the here and now. It is noticeable how little the media attempt to make their own environmental agenda. Thus their two periods of greatest concern – in 1987–9 over ozone layer depletion and global warming and, far more briefly, in May and June 1992 just before and during the United Nations Conference on Environment and Development (UNCED) – largely reflected the expressed concerns of politicians and world leaders. When these subsequently switched their attention elsewhere the media tamely followed. Global warming and ozone layer depletion are still of course with us, still represent as serious a threat as when they were first 'discovered' in the late 1980s. But they are no longer exciting, no longer make the front page and have been demoted to just two more of a long list of problems with which we have to live.

Recent research by Lacey and Longman (1993, 1995) has underlined the extent to which British media coverage on the environment also reflects broader political preoccupations and shows that the more embarrassing specific environmental issues are to the government, the more likely it is that the majority of the press will keep silent or be hostile to green perspectives. Equally fundamentally, it confirms that media coverage on environmental issues largely reflects the government's agenda and shows little consistency over time.

Environmental coverage is still largely confined to its own ghetto. Closely related political, economic, and security issues continue to be treated separately. The strong preference is for concrete, preferably photogenic, issues such as road building and other threats to habitats at the expense of broader and longer-term analysis. Apart from the occasional moan at meetings of environmentalists and the odd twinge in green magazines, this problem is rarely raised in public. But it is central to the failure of the ecological movement to challenge the continuing dominance of the sectoral, short-term, and pragmatic in British political perceptions.

THE ROLE OF SCIENTIFIC UNCERTAINTY

I turn now to the scientific basis of policymaking. It is after all scientifically based fears of environmental degradation that have driven much of the debate. For reasons which I have no space to explore in this chapter, those fears have largely centred around global warming. It has largely been fear of melting Antarctic ice-caps, of rapidly rising sea levels swamping Bangladesh and New York alike, of temperature and rainfall changes turning fertile lands into deserts, which has driven pressures for dramatic changes in our policies, aspirations, and lifestyles and led us to the extraordinary situation where carbon emissions from our power stations and methane emissions from our cows and rice paddies are treated as deadly threats to human survival.

Yet it is precisely in the field of global warming that time-scales impose the greatest uncertainties. The key problem is this. Climate patterns have fluctuated considerably in the course of this century. But there is so far no evidence that these variations have been other than naturally caused. The 'evidence' of global warming is so far entirely confined to the global climate models (GCMs) which are put together by scientific teams. The scientists involved are the first to recognise the weakness of these models. As the physics has improved, as the oceans and cloud effects are more effectively incorporated, assessments of the magnitude of the change in store have come down. For example, on current projections sea-level rise at 4 cm per decade will be 24 cm by 2050 – far less than thought likely two to three years ago. Similarly, estimates of changes in rainfall patterns and temperature increase have been revised downwards.

We are unlikely to see evidence of climate change emerging for another ten years or so. Nor are we likely to know earlier what the regional impacts will be. Obviously the figures could also be revised upwards as models improve. It would be rash to make any forecast.[2] My point is rather that there is a mass of uncertainty which is unlikely to be resolved in the near future and that this is likely to diminish the credibility of global warming as a front-runner in the global ecological threat stakes. But global warming is only one symptom of the ways in which human economic activity is remorselessly damaging the natural ecosystems on which we depend for our survival. There are far fewer, and far less serious, problems of time-scales and causality in almost all other areas of serious concern. Strong scientific evidence that anthropogenic change is increasingly overwhelming 'natural' change and is reducing the ability of natural systems to absorb the consequences of human activities is abundant, for example, in terms of soil erosion, acid deposition, depletion of fish stocks, deforestation, and ozone layer depletion.

This is not of course to argue that, global warming apart, there are not many gaps in our knowledge of what is happening to the environment, whether at global, regional, national, or local level. As even a cursory reading of the key literature will confirm, these exist (e.g. Tolba and El-Kholy 1992; World Resources Institute 1994), notably in terms of data for developing countries. My contention is rather that such information and analysis as we have provides more than enough evidence for the policymaker that in almost all areas of ecological concern and in almost all regions of the world we are going fast in the wrong direction and that we need to take urgent corrective action.[3] Uncertainties resulting from relatively long time-scales of natural environmental change, whether or not linked to adequate understanding of natural fluctuations, are exciting to the academic. In the 'real' world of politics and government they may provide a convenient excuse to the policymaker to avoid taking action. But very rarely will the excuse be valid, or the real reason for delay.

CONCLUSION

One of the central themes of the Brundtland Commission Report is the clash between 'political' reality, or what political leaders feel they can achieve inside existing political and economic parameters, and 'ecological' reality, or what actually needs to be done to save the environment. Rifkin (1987, p. 237) rephrased this in temporal terms, as a conflict between 'artificial' and 'natural' time. 'Balancing our budget with nature,' he argued, 'requires that we reorient the pace of our economic activity so that it is compatible with nature's timetables.'

This need to reconcile human and natural time-scales has important implications for our political system. The time-scale of energy policy, for example, will need to be reconceived over a period of at least sixty to one hundred years if the world's rapidly growing population is to meet its basic needs for energy services while slowing down global warming, reducing acid deposition and other, more 'local', forms of pollution, hoarding our remaining supplies of fossil fuels so far as possible and moving over to energy systems based on renewables. Moving from our current system of basing taxation on 'goods', such as employment and investment, to a fully fledged system of 'eco-taxation' based on taxing 'bads', such as waste and energy use, would in the views of the key proponents of such a system, such as Professor Ernst Ulrich von Weizacher, take perhaps fifty years. In the area of regulation and standards business leaders such as John Wybrew of Shell argue for a similarly long period of close cooperation between government and industry to promote ever higher standards of resource use and pollution control (the process of 'de-materializing production').

But the need is not just for longer-term policies. Perhaps even more radically, the reconciliation of time-scales demands an end to the basically confrontational and see-saw nature of British politics, under which an incoming government routinely reverses the key policies of its predecessor. Such a system is incompatible with the pursuit of the long-term national consensus which alone will enable long-term policies to be argued out, adopted, implemented, and maintained despite changes of government. Instead, building agreement on where the long-term national interest lies and how it might best be pursued will need to be at the heart of politics. The difficulties are obvious. The only possible starting point is widespread recognition of the need for such fundamental change. It is striking that such recognition is almost entirely absent from the current otherwise lively debate about Britain's political future. Our system needs, and lacks, 'time guardians' to remind us.

NOTES

1 In the first example, assessment is based on the author's experience when in Whitehall, in the second on his subsequent experience in promoting dialogue at the Global Environment Research Centre between government and NGOs on international trade issues.

2 I am grateful for help with this section to Sir John Mason, FRS, my former colleague at the Global Environment Research Centre.

3 This was of course the main theme of the Brundtland Report and, implicitly, of Rio itself, and as such is hardly controversial.

REFERENCES

Lacey, C., and Longman, D. 1993. The press and public access to the environmental debate. *Sociological Review* 41(2): 207–24.

Lacey, C., and Longman, D. 1995. Modelling the U.K. Press's Coverage of an International Event: the Press and Cultures of Understanding, *MEEG Memo* No. 1 (September).

Rifkin, J. 1987. *Time Wars*. New York: Simon and Schuster.

Tolba, M. K., and El-Kholy, O. A. (eds) 1992. *The World Environment 1972–92*. London: Chapman and Hall for UNEP.

World Resources Institute. 1994. *World Resources 1994–95*. Oxford: Oxford University Press.

8

ENVIRONMENTAL DESTRUCTION IN SOUTHERN AFRICA

Soil erosion, animals, and pastures over the longer term

William Beinart

Editors' note With this chapter we shift the scale and focus of attention. William Beinart is concerned not with climatic change *per se*, nor with a global perspective. He is concerned with a case study of how during this century different actors – government officials, scientists, white farmers, black herders, apartheid politicians, and post-apartheid politicians – have interpreted environmental change in the rangelands of the Karoo of South Africa. This is the human lifetime scale at which some people believe they can detect environmental degradation, and seek to find the causes. But as Beinart points out, the evidence is very contradictory, and there are unspoken assumptions about what is right or good in the first place – mostly that, despite the use of the rangelands for stocking, somehow they should remain the same, in some undefined equilibrium, and further that white settlement meant progress, even if it had a few managerial hiccoughs along the way. The Karoo also happens to be the kind of area about which Neil Roberts wrote in Chapter 2: low latitudes where there could be great swings in precipitation and where landscapes might be far from biological and geomorphological equilibrium. Beinart's view that what we should be concerned with is change and the management of change should chime with both Jean Grove's and Max Wallis's views, expressed respectively in Chapters 3 and 5.

INTRODUCTION

A great deal of the literature written on South Africa which explicitly addresses environmental issues paints a picture of decay and degradation over the long term. One of the most persistent concerns, and an important focus for commentators, has been soil erosion (Beinart 1984). In a country where stock were important both for white and for black farmers, over-stocking and overgrazing have frequently been cited as a major cause of denudation, desiccation, and erosion. The state of the settler stock farms and semi-arid areas more generally has been the trigger for broader debates and discourses about ecological decay in the region. This chapter examines ideas about degradation on white-owned stock farms in the lower-rainfall grazing regions of South Africa – more than half the area of the country – in the light of some fascinating recent research in botanical history, in particular the work of Hoffman and Cowling.

Hoffman and Cowling are uneasy about the picture so frequently painted of continuous degradation in natural veld pastures in the Karoo and its better-watered grassland peripheries. To some degree their findings echo an earlier, but minority, botanical viewpoint (e.g. Tidmarsh 1952). They tend to explain their less alarmist approach by emphasising that the vegetation of this area has been subject to fluctuation shaped largely by natural causes. This chapter offers some other tentative explanations as to why, from a historian's point of view, their position may have some foundation.

My approach to these questions is cautious not only because it is difficult for a non-specialist to interpret complex and sometimes conflicting scientific data, but also because my work is far from complete. However, the argument is intended to be consistent with two very general points which do not in themselves depend upon the material analysed here. First, while there is no doubt that growing population, stock numbers, commercialisation, and apartheid have contributed to environmental problems in Southern Africa, environmental historians are on shaky ground if they stalk the past only with the limiting concepts of decay and degradation. Environmental change should be examined in a less linear manner; deployment of the concept of transformation, rather than just degradation, might help to shift the emphasis of debate. Second, it may be unwise to use arguments, canvassed in recent years, about the environmental destructiveness of settler capitalist agriculture in order to justify totally new ownership regimes in South Africa.

THE STOCKING OF SOUTH AFRICA'S RANGELANDS

South Africa's Karoo, together with its better-watered eastern fringes and the southern high-veld grasslands of the Free State, has been browsed and grazed for a very long time by a large variety of indigenous species of wildlife. Before settlers conquered and took control of the region in the eighteenth and early nineteenth centuries, parts had been used for many centuries by the flocks and herds of the Khoikhoi. African people settled on the east coast also used the grassland fringes seasonally. It is therefore difficult to argue for a pristine period when vegetation existed in a natural state, unaffected by animals and people.

The initial expansion of the settler grazing frontier was driven by the market for meat. Settlers ran cattle and the fat-tailed hairy sheep which they had adopted from the Khoikhoi, but they used the veld more intensively. The myth of the subsistence-oriented South African trekboer has long been disputed by historians; as with people on other settler colonial frontiers their mode of life could be rapacious. One indication of this is that as early as the 1830s, much of the wildlife had been shot out of the more heavily stocked districts both to secure meat supplies and to reduce competition for grazing. By then, two species, the quagga and bluebuck, were extinct in the wild. Imported technology (notably guns and horses), together with social disruption in this early colonial period, simultaneously brought new communities of Griqua (regrouping slaves and Khoikhoi) and Africans into the semi-arid interior.

From the 1830s woolled sheep, mainly merino, were absorbed into a funnel of land between the 250 and 630mm rainfall lines, spilling over into the wetter east coast districts. In the second half of the nineteenth century these thinly populated districts were major growth areas in the Cape Colony and Orange Free State. Although the value of wool exports was overtaken by that of diamonds and then dwarfed by that of gold, wool remained the most important agricultural export well into the twentieth century. The figures on sheep holding show extraordinary growth in South Africa. It is likely that there were fewer than 4 million sheep in 1840, 10 million by 1865 and close to 25 million by the 1891 census; additionally, fat-tailed sheep, angoras, and other goats were numerous in particular districts. Numbers declined during the South African War (1899–1902) but then grew rapidly to peak in 1913 at about 28 million woolled sheep and 47 million small stock. In 1931, they reached an all-time high of about 45 million woolled sheep and 58 million small stock in all (Figure 8.1).

While the areas discussed in this chapter were largely sheep districts, some also carried cattle. Cattle numbers fluctuated more wildly around the turn of the century because of diseases such as rinderpest (1896–7) and East Coast fever (1904–13), as well as the South African War. Nevertheless, the overall trend was rapidly upwards from under 4 million in 1904 to about 8 million in 1920 and over 12 million by 1939. The most spectacular growth was in the Orange Free State and Transvaal, but even in the Cape numbers doubled over less than four decades. Whereas the great majority of sheep were owned by white farmers, nearly half the cattle in the country were owned by blacks as late as 1930. There were, of course, many other types of stock – horses, mules, donkeys, ostriches – but these were mostly more dependent on fodder and are not as significant for the argument as sheep and cattle.

The control of new animal diseases such as scab, redwater, rinderpest, East Coast fever, and heartwater – a major preoccupation for colonial Agriculture Departments – facilitated rapid expansion in numbers. Near-universal dipping of sheep and cattle was achieved by about 1920.

Water was also a major constraint on pastoral farming. The original location, size and layout of farms was shaped not least by the availability of

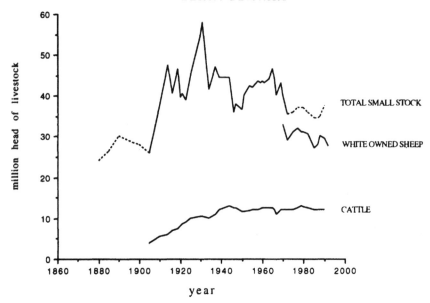

Figure 8.1 Cattle and small-stock numbers in South Africa

Note: Figure 8.1 must be seen as representing approximate numbers. It attempts to include all the stock in South Africa, whether owned by black or white. The original figures in various censuses and agricultural censuses are somewhat suspect and numbers are especially weak since the 1970s as statistics for the black homelands are decreasingly reported in national figures. For example, there has probably been a significant increase in goat numbers in the homelands which is not fully refected in available figures. An attempt has been made to present total figures, rather than to segregate out those for whites only, for three reasons. First, at least till the 1950s, many of the stock on white-owned farms belonged to black tenants and workers. Second, totals also allow for local boundary changes: the reserve or homeland areas have approximately doubled in size since the early twentieth century and the white-owned farmlands have declined in extent. Third, environmental impacts, while they can be very localised, often spill over political boundaries.

natural water sources such as springs, vleis, and streams. Dam building was one alternative means of controlling supplies, and from the mid-nineteenth century shallow earth farm dams began to pepper the countryside. Boreholes – drilled by the Department of Irrigation as well as private firms – increasingly became an alternative as technology improved in the early twentieth century. They tended to be more reliable in that they tapped underground sources which were relatively constant. Metal windmills sprouted across the dry regions of the country. South Africa did not quite become a new 'hydraulic society' after Worster's model of the American west (Worster 1985), but rather, in stock-keeping country, a borehole society. In the early decades of the century, farmers and the state acted in concert to eliminate predators of small stock, especially jackals and caracals; hundreds of thousands were killed in response to offers of high bounty payments.

ASSESSMENTS OF ENVIRONMENTAL IMPACTS ON THE RANGELANDS

The cost of such rapid expansion of animals, observers increasingly argued, was to the state of the veld. There is a long record of colonial comment on its deterioration, dating back to the period of Dutch East India Company rule before 1806. Some of these were collated by T. D. Hall, perhaps the first to attempt a systematic history of pastures, in the 1930s and 1940s (Hall 1934, 1942). An English-speaking South African (b. 1890), brought up in the Free State and trained in the United States, he became agricultural adviser to the leading fertiliser company in the country. After extensive research into soil chemistry and veld management, he became a strong conservationist and this led him to explore the historical sources. He found officials near Cape Town commenting in 1751 on 'the disappearance of grass and the springing up of small bushy plants in its stead'. In 1775 the Swedish traveller Sparrman noted that

> in consequence of the fields being thus continually grazed off and the great increase of the cattle feeding on them, the grasses and herbs which these animals most covet are prevented continually ... from thriving and taking root, while on the contrary the rhinoceros bush which the cattle always pass by ... is suffered to take root free and unmolested.
>
> (Hall 1934, p. 66)

Such observations were in turn linked to wider developments in European thought, which had long recognised the way in which natural resources were disturbed by agricultural development and intensification (Glacken 1967). Richard Grove (1994) has illustrated the extent of Dutch environmental concern about the new Dutch colonies as early as the seventeenth century. He suggests that colonial experiences, especially in transforming the ecology of small islands, fed back as a major strand into developing Western environmental thought. By the mid-nineteenth century, following the burgeoning of Enlightenment science, many of the linkages which currently inform environmental thinking had been made – in particular those between denudation and soil erosion. Desiccationist ideas – which connected reduced vegetation with climate change, declining rainfall, increased evaporation, and drought – were influential.

John Croumbie Brown, the innovative Scots Colonial Botanist at the Cape in the 1860s, wrote extensively and alarmingly on these issues and proposed far-ranging solutions (see Grove 1987). By 1873, a short report by John Shaw to the British Association for the Advancement of Science outlined the key points which were to become commonplace over the next century:

> After sketching the distribution of plants in South Africa, the author went on to particularize the character of the prairie-like midlands of the

Cape with their luxuriant grass and vegetation. Since sheep have been introduced the grass has fast disappeared, the ground (by the hurried march of sheep for food amongst a scattered bush) has become beaten and hardened, and the seasonable rains which do come are accordingly allowed to run off the surface without soaking into the ground to the extent formerly the case. The country is thus drying up, the fountains becoming smaller and smaller, and the prospect is very clear that the midland regions will turn into a semi-desert. Indeed the plants of the singular regions known as the Karoo, in the south-west of the Cape ... are travelling northwards rapidly and occupying this now similar dry tract of country. The herbage is essentially a Karoo one already.

(Shaw 1873, p. 105; see also Cape of Good Hope 1877)

T. D. Hall was only one of a number of people addressing the problem between the 1910s and 1930s at the time when the growth of stock numbers was particularly alarming (Figure 8.1) A Senate select committee of 1914 produced an influential report on droughts, rainfall and soil erosion (Union of South Africa, Senate 1914). Afrikaner intellectuals, some of whom had also trained in the United States, were at the forefront of concern. Botanical and environmental knowledge was one of the strong intellectual traditions in South Africa, rooted in the rich travel literature as well as late-nineteenth-century scientific professionalisation and in experiences on the land.

Enslin, the chief Inspector of Sheep in the Department of Agriculture, laid particular stress on the damage done to the veld by 'tramping' – both the seasonal treks of animals in long-distance transhumance and daily movements to widely dispersed water sources on farms. Seasonal burning of veld attracted widespread criticism. The Drought Investigation Commission report of 1922/3 (Union of South Africa 1923), which concentrated on stock farming, took extensive evidence and brought together many previously articulated anxieties in language which was powerful and persuasive. Overstocking, denudation, and drought threatened a 'newly-created South African desert'. The old-established practice of 'kraaling' sheep at night, or bringing them back to a central byre in order to protect them from theft and predators, was particularly criticised.

This environmental alarm, and the understanding that South African pastures were overstocked and had been in continual decline over a long period, was firmly established in the inter-war years. A detailed history of shifting emphases in the debate has yet to be constructed. Earlier commentators seem to have emphasised changes in vegetation, while climate change, overstocking, and soil erosion later became more prominent concerns. Anxiety was perhaps first most acute in respect of extensive white-owned farms. But it was frequently articulated in the nineteenth century, especially in connection with deforestation and burning, about African societies. By the time of the Native Economic Commission (Union of South Africa 1932), the basic points about environmental decline in the reserves were firmly entrenched and have permeated a great deal of literature ever since.

Significant legislative action was taken both in white-owned and black-occupied districts; soil conservation became a part of post-Second World War reconstruction plans. A further inquiry by the ominously named Desert Encroachment Committee (Union of South Africa 1951) aimed to examine more definitively whether 'man-made desiccation had altered the natural condition of the veld to such an extent that the climate itself had in turn been affected' (p. 2). The relationship between rainfall and denudation, extensively researched in Europe and the United States, had long been debated in South Africa. Croumbie Brown in the nineteenth century, and many since him, argued that there was a strong correlation. As the Committee reported, 'at all times there appears to have been a fairly widespread belief that climatic conditions were slowly, but progressively deteriorating' (p. 1).

As in the case of the Drought Investigation Commission thirty years earlier, the report was sceptical about any long-term decline in rainfall. Rainfall records suggested that there had been a peak around 1890, and a gradual, irregular fall until the droughts of the early 1930s. But subsequently rainfall had been relatively high, not least immediately before the Committee reported. Committee members accepted that there might be such long-term cycles although these were difficult to detect with the evidence they had available. (This cycle is now more widely accepted and there is clearly some form of shorter-term seven- or eight-year cycle on the east coast of the country – even if it is not always regular.)

On the issue of veld deterioration, however, the committee expressed no such scepticism. Most of the evidence was again taken in and around the Karoo. The report was strongly influenced by the work of John Acocks, an ecologist who had worked for the Department of Agriculture since 1935 and conducted extensive fieldwork. He was completing his own national survey, *Veld Types of South Africa* (first published a little later in 1953; Acocks 1975), and he also served on the Desert Encroachment Committee. As in the case of the Drought Commission, Acocks's work fixed the language of debate and is still the mostly widely quoted source on pastures and veld types. Although Acocks did not devote an extensive section to pasture history, the authority of his work was such that his observations have become commonplace.

Acocks noted the general desiccation of the Karoo. In its drier western parts it had turned into 'near desert in the sense that soil erosion is universal and that there is no longer a permanent, unbroken vegetation cover, and only rarely a temporary cover' (Acocks 1975, p. 8). Most strikingly, he reiterated the points made by Shaw, and others since, that Karoo vegetation had moved eastwards, in parts by up to 250 km, to replace sweet grassveld. There was also a slower, but accelerating, northward movement.

Elsewhere, sour or rank grassveld, less suitable for year-round grazing, had spread at the cost of mixed and sweeter velds, especially the climax grasses, notably *Themeda triandra* or *rooigras*. In short, there was 'widespread deterioration in all veld types over the last 500 years' (Acocks 1975, p. 78). Acocks was particularly concerned about the development of what he called the False Upper Karoo: 'The development of this veld type constitutes the most spectacular of all the changes in the vegetation of South Africa. The conversion of 32200 square km of grassveld into eroded Karoo can only be regarded as a national disaster' (ibid.).

The Desert Encroachment Committee incorporated these ideas and reiterated a view often expressed in the inter-war years that the state should take a lead in resolving the crisis: 'it is quite clear that farmers do not know what is best' (Union of South Africa 1951, p. 7). Members also examined research done on siltation of dams and, less conclusively, on possible depletion of underground water as a result of the rapidly increasing number of boreholes:

> It would not be too much to say that the determination of correct systems of veld management combined with the optimum stocking rate for every veld, soil and climatic zone would be one of the greatest contributions that this generation could make to the future welfare of agriculture in South Africa.
>
> (ibid.)

In language very reminiscent of the Drought Investigation Commission's report, and of American inter-war commentators such as Paul B. Sears, the Committee concluded that 'the very existence of stock farming is at stake because, unless we can succeed in arresting further deterioration of the veld, it is doomed'. (By this time sheep farming in particular was far less central in the national economy.)

Acocks spent some of his career based at Grootfontein, the major agricultural research station and training college in the Karoo founded in 1911 at Middelburg. P. W. Roux developed further research at Grootfontein over a long period from the 1960s and confirmed many of the tendencies highlighted by Acocks. He suggested a five-phase progression of deterioration from pristine veld to desert. Much of the Karoo was in the third phase; woody invader plants were still spreading 'at an alarming rate'.

It is important to note that most of the authors of the material cited in the twentieth century – and a good deal more could be mentioned – were progressive farmers, scientists, and government officials who were generally not antagonistic to the broader segregationist policies being pursued by successive governments. Yet this powerful and well-established technical thinking was readily incorporated into a far more critical anti-apartheid literature on the rural areas. Aside from the fact that physical evidence of decay like dongas (gulleys) could be observed, it is not difficult to understand why. On the one hand, ideas about degradation provided an argument with which to castigate the greed of white farmers who were at the core of support for apartheid and were, together with the state, seen as responsible for severe rural dislocation through forced removals. On the other hand, they helped to illustrate the iniquities of the homeland system in which blacks were restricted, as independent occupiers, to a very limited proportion of the country's land. Whereas the Native Economic Commission had blamed African culture and attitudes for ecological degradation in the reserves, it was not difficult to invert the argument and pin the responsibility on the restrictive policies of apartheid.

Many of the more general publications dealing with apartheid and environmental decay correctly concentrate on the way in which severe environmental ills have been concentrated in the homelands. Some also summarise key points in regard to white-owned farms. *Uprooting Poverty* (Wilson and Ramphele 1989) and *Restoring the Land* (Ramphele 1991), two important books by anti-apartheid social scientists, which deal with poverty, ecology, and post-apartheid reconstruction, are good examples. Despite government measures, they argue, 'the rape of the soil has continued'. Francis Wilson quotes the estimate that 400 million tonnes of topsoil was being lost annually in the 1960s. Overstocking on sheep farms in the Karoo and Cape Midlands was around 30 per cent and the desert was advancing at about 2.6km per year.[1] (This would suggest it had spread 100km further since the Desert Encroachment report (Union of South Africa 1951).) He reiterates Acocks's view that over the centuries sweetveld had become scrub and that by 2050 the sheep-farming areas would be desert. 'The combination of soil erosion and deteriorating vegetation is held by some to be far and away the most serious of the many problems facing South Africa' (Wilson in Ramphele 1991, p. 30).

Alan Durning's hard-hitting Worldwatch pamphlet, *Apartheid's Environmental Toll* (1990), also mentions some of these points; he actually talks of the 'south-western deserts ... marching to Pretoria, expanding across two and a half kilometres of exhausted pastures a year' (p. 6). (In his echo of a South African War song, he omits to mention that the Karoo would have to cross over a few hundred kilometres of the better-watered highveld crop belt if it were really to get to Pretoria.) Huntley (a leading South African environmental scientist) and co-workers (Huntley *et al.* 1989) see the 1980s as 'the decade the environment hit back for over a century of careless environmental management'. They do not accept a doomsday scenario where resources are finite and population growth of necessity a route to disaster. Nevertheless, they do accept the idea of continuous decline manifest in the unparalleled diversity and seriousness of natural disasters – floods, droughts, urban degradation, hail, fire, locusts – which they argue were worse even than those in the 1930s. 'We have had to pay the price for over a century of careless environmental management and South Africa's unique experiment in social engineering' (p. 37). They similarly estimate that annual losses of soil are 300–400 million tonnes, nearly 3 tonnes per hectare, and that the Karoo is still overstocked by almost 50 per cent (p. 39). Overall, they argue, 3 million hectares have become unusable because of erosion. They also reflect increasing concern, manifest at least since the mid-nineteenth century, about biodiversity.

The botanist Richard Cowling, speaking in a slightly different voice,[2] notes that a particularly large number of species are at risk because of the unusually diverse and unique nature of Cape plants and animals.

> The most important cost associated with agricultural development in Southern Africa has been the degradation of natural resources and ecosystems.... The erosion of topsoil, which is practically a non-renewable resource, is unacceptably high .... Most of the natural grazing land is seriously overstocked and as much as 60 per cent of the veld is currently in poor condition.
>
> (Cowling 1991, p. 16)

Most of these authors are reluctant to blame capitalism itself. However, Cowling, approaching the issues from an apparently innocent ecological viewpoint, comes up with some very radical suggestions. Because it is wrong, he argues, to confine stock to a small area in semi-arid zones, 'an ecologically appropriate intervention in the Karoo might involve the removal of barriers to stock migration and nationalisation of the herd' (ibid., p. 17). Taken to its logical conclusion, this analysis might suggest that 'ecological conditions would favour communal land tenure and nomadic pastoralism'. I shall return to this point in my conclusion.

DEGRADATION OR TRANSFORMATION

It is not my intention to argue that the idea of continuous degradation amounted to a unquestioned 'conventional wisdom' which had no foundation. In the first place, many of the protagonists of this view regarded themselves as somewhat unconventional in the face of the weight of a powerful farming lobby or, later, an uncaring government bent on its programme of massive social engineering. Second, some of the observations about ecological decay undoubtedly ring true. Certainly for the period up to the early 1930s, the growth of stock numbers, especially given the methods of stock farming practised, must have had major and far-reaching effects. While it is difficult to make cross-national comparisons because of the variety of ecological conditions, South Africa does appear to have been comparatively heavily stocked at the time. Only Australia, India, the Soviet Union, the United States, and possibly China and Argentina, all far bigger countries, had more woolled sheep. (Countries such as New Zealand and the UK had far higher numbers per hectare but are of course much wetter.)

However, I do want to raise questions about the linearity and character of change and the concept of degradation itself. It is striking that a continual buildup of stock numbers was possible for many decades after alarmist reports in the nineteenth century. And after apparently devastating levels of stocking around 1930/31, and calamitous losses in the early 1930s, numbers of both sheep and cattle rose quite sharply following better rainfall in the mid-1930s. Similarly, after a further fall in stock numbers, twenty years more of reported desertification, and the dire warnings of Acocks and the Desert Encroachment Committee, small-stock numbers again rose sharply in the 1950s (Figure 8.1).

One reason for the resilience of national small-stock numbers up to the early 1960s may of course be that increases took place in newly stocked districts, while areas which had been heavily stocked in the nineteenth century experienced more severe losses. It is true that sheep were increasingly incorporated into the agricultural systems of parts of Natal, the northern Orange Free State, and the Transvaal in the early decades of the twentieth century. And according to the Drought Commission of 1923, degradation was already leading to a fall-off in numbers in the old sheep districts of the Karoo and midland Cape – some of them well stocked for over a century.

A definitive response to this contention would require careful district by district analysis of stock numbers over a long period. But it does not seem to hold in any simple sense for the first half of this century. For example, sheep numbers increased quickly in the Cape Province during the 1920s, immediately after the Drought Investigation Commission reported that they were undergoing a long-term decline (Union of South Africa 1923). Moreover, losses both in sheep numbers and in wool production in the great 1930s drought were proportionately worse in the Orange Free State, Natal, and Transvaal than the Cape – although the Cape had been stocked for a longer period.

The statistical evidence for the years up to the 1950s certainly requires further investigation. But I should like briefly to explore different approaches to the question, which do not necessarily contest the view of environmental degradation at some periods, but raise question marks about a continuous long-term decline.

One of the botanists serving on the Desert Encroachment Committee was C. E. Tidmarsh, who also worked at Grootfontein College of Agriculture. To a greater extent than other experts (Acocks shared some of his views), he was cautious about a linear view of decline. In a short article written in 1952, Tidmarsh reported on some years of veld experiments. He agreed that knowledge was still inadequate but noted:

The results of the past seventeen years of research at the College have shown clearly that the amount of natural vegetation that can be maintained per morgen of land, is controlled more by the available moisture supply than by the grazing treatment to which the veld may be subjected, and that in the extensive flats of the Mixed Karoo, the quantity of vegetation growing at present on the soil is, with the exception of local areas of denuded soil, in approximate equilibrium with the available moisture supply, and that, without increasing the latter, it is virtually impossible to increase the natural cover of the soil by any measure of grazing control, including complete protection.

(Tidmarsh 1952, p. 1)

Even more controversially, Tidmarsh argued that while extremely heavy stocking rates could change the composition and quality of the veld, it was very difficult to produce a lasting impact. 'Within limits, the composition of the veld appears, thus, to be more a function of the interaction of the soil type, moisture supply, and climate, than of the grazing treatment' (ibid., p. 1). Noting similar results in the United States, he tentatively suggested that continuous grazing at a moderate rate of stocking was not necessarily harmful. It is interesting that Tidmarsh's contention does not contradict the findings of an American agricultural economist who reported on South Africa during the 1930s Depression (Taylor 1935). He calculated that the number of sheep per acre in any district correlated strongly with average rainfall. In some ways these views echo the arguments of the American pasture historian J. C. Malin, who de-emphasised the role of people, and particularly European settlers, in shaping pasture histories over the long term in the United States (Worster 1979).

In the post-Depression United States, where soil conservation was one major justification for state intervention, Malin's views were associated with an anti-federal and conservative anti-interventionist political viewpoint. For Tidmarsh, Acocks, and most other South African experts, by contrast, the role of the state was unproblematic – the problem was more how to intervene effectively. Tidmarsh did recognise that there were limits to safe stocking. Both he and Acocks argued that whatever the long-term effects of pasture use, selective grazing of more palatable species could have serious short-term consequences. It had long been argued in South Africa that grazing in fenced paddocks, rotated through the year, increased the carrying capacity of the veld. They refined experiments on systems of rotation and generally supported more intense use of smaller paddocks, rotated frequently in order to minimise selective grazing.

Echoes of Tidmarsh's approach can be found in botanical work by Hoffmann, who is cautiously beginning to contest the view that Karoo vegetation has been extensively altered and is spreading (Hoffmann and Cowling 1990). First, together with Cowling, he has assembled a formidable range of historical sources on the appearance of the Karoo before the period of intensive farming of woolled sheep. They argue, with reference to the eastern Karoo, that 'although there are some references to a grass-dominated landscape, even the earliest accounts suggest that, at least in places, dwarf karroid shrubs were dominant' (p. 289).

Second, and more dramatically, they have been involved in systematic photographic research. In view of the long concern about the state of the veld, numerous photographic records taken by botanists survive. These include a series by Pole-Evans, the Cambridge-trained botanist who came to work as a government scientist in South Africa in 1905 and emerged as one of the most influential figures in stimulating research on grazing problems and conservation more generally (Scoones 1994). Matched photographs taken in 1989 suggest that the state of the veld had improved considerably since 1917–25 when the earlier photographs were taken.

Third, Hoffman and Cowling resurveyed eleven sites in the Karoo and southern Orange Free State which had previously been investigated by Roux in 1961–3. Roux, who became head of Grootfontein, participated in a sample survey in 1989 so that better comparability could be achieved. 'All sites showed an increase in total percentage canopy spread cover from 1961–3 to 1989, attributed chiefly to an increase in the cover of grasses', they found (Hoffman and Cowling 1990, p. 290). Some sites showed a decline in shrub cover.

On the basis of these findings, Hoffman and Cowling suggest that there is not strong evidence for an expanding Karoo. Rather they see the possibility of short-term changes in response to rainfall. They did not think that the prevalence of grass cover recorded in 1989 constituted evidence of a reversal of processes noted earlier and that grasses were now invading the Karoo. This phenomenon was also more likely to be part of shorter-term cyclical change, possibly due to higher rainfall in the mid- to late 1980s. 'Except for a very general understanding, we do not know what influence grazing has on these processes' (Hoffman and Cowling 1990, p. 292).

It is very difficult for a historian untrained in botany to evaluate these various studies, but it is possible to put them against other evidence – notably long-term stocking rates and other factors governing the intensity of veld use. Figure 8.1 gives a rough sense of national trends, (although see the note on p. 152 about its problems). Pole-Evans's pictures and the others used for the period 1917–25 were taken after a decade of spectacular growth in small-stock numbers, from about 26 million to 47 million – and when numbers reached their early peaks. Cattle numbers were also increasing rapidly, though not necessarily in the sheep districts. Given that stock were grazed almost entirely on the open veld at the time, and that paddocks were not universal, it is unsurprising that the pictures present a denuded landscape. Surveys done in 1962–3 by Roux and his colleagues similarly took place at a time of high small-stock numbers, after more than a decade of sustained increase, and near to their post-Second World War peak.

By contrast, 1989 was at the end of a period in which small-stock numbers had been low for over fifteen years. Indeed, from the early 1970s until the late 1980s, small-stock numbers in the country as a whole were lower than at any time since the first decade of the century. Cattle numbers were also fairly stable in this period. If the figures for white-owned sheep alone are taken – these make up the great majority of farm animals in the districts being discussed – they may show an even more significant fall in the 1980s. Although they rose a little in the late 1980s, they have remained relatively low since then because of the collapse of the wool price and serious drought in the early 1990s.

If this evidence about small-stock numbers is put together with the claims by Tidmarsh and others about the capacity for veld to recover under favourable circumstances, it is not entirely surprising that Hoffman and Cowling's pictures showed improvements. Two further related issues should be raised: first, whether there are other factors which might have led to improvement in veld; and second, whether stock numbers have remained low simply because the veld could carry no more.

Numbers are by no means the only factor which might affect pastures. On the one hand, the size and eating capacity of sheep have probably increased in the period under discussion; 35 million small stock in the 1980s might eat as much as 45 million in the 1910s. On the other, there has been widespread investment in grazing management. The Drought Investigation Commission report (Union of South Africa 1923) predicted that the carrying capacity of veld could increase by 33–75 per cent by the use of camps and that the output of wool could double. Internal divisions on sheep farms go back a long way. Despite the size of the farms, dry stone walls were built in the nineteenth century and these can still be seen in some of the older, and formerly wealthier, sheep districts. There was certainly no shortage of stones in the Karoo. Barbed wire replaced stones from the 1880s, and was increasingly important in keeping jackals out of paddocks as well as keeping sheep in. The expense of fencing was diminished by state subsidies in the early decades of this century.

T. D. Hall noted that Graaff-Reinet farmers, in the heart of the old sheep country, had generally fenced by the time of the 1930s droughts. Although they farmed in an area with only about 380mm of rainfall, they experienced fewer losses than those in many other sheep districts. In subsequent years, especially in the period of high commodity prices, high subsidies, and the wool bonanza of the early 1950s, many large sheep farms were more systematically subdivided into fenced paddocks. Extension officers provided farm planning advice, and veld types could be more systematically identified as the basis for the division of grazing camps.

One inhibiting factor on the multiplication of camps was the difficulty of water provision in all of them. There were not always suitable borehole and dam sites. The use of cheaper plastic piping to distribute water around farms from the 1970s was of major importance in resolving this difficulty and has also facilitated further fencing. Water could now be brought to camps, rather than camps extended to water. Interviews on sheep farms, especially those which have stayed in the same family over a few generations, suggest that internal fencing has been a continuous process over many decades. Fences represent a very considerable element in the value of farms, which are seldom below 1,000 ha and usually much bigger in the dryer districts. Farmers like to comment that given the low price of land, the cost of fencing from scratch would now be the same as the price of land. There has been and remains intense debate about the most appropriate forms of grazing, size of camps, and frequency of rotation. But it seems very likely that systems of rotation have been one factor in increasing yields, diminishing selective grazing and reducing the effects of tramping by keeping animals in one place.

Some fodder was developed for sheep, particularly prickly pear and agave from Central America. Both were used, chopped, as drought foods. Spineless versions of prickly pear were bred in the early decades of this century. Lucerne spread during the ostrich boom of the late nineteenth and early twentieth centuries, and is widely used – for example in raising angoras – wherever sufficient water can be found. In some areas, lambing is done on small areas of irrigated green fodder such as oats, which is also used as winter feed. During the 1930s, in both the United States and South Africa, experiments were done with the best indigenous grasses for planted pastures. Their development has been slow but subsidy programmes in the 1980s have greatly hastened their spread. Various other factors, such as the elimination of white and black tenants, largely in the first half of this century, and the increase in average farm size, which roughly doubled between the early 1950s and mid-1980s, have probably also have contributed to veld stabilisation.

While there is no doubt that stock numbers have declined, there is scope for debate as to the causes of this decline and hence its implications for the state of the veld. As noted above, the Drought Commissioners (Union of South Africa 1923) argued that both the sheep and the human populations (they referred to whites only) of some of the key old sheep districts were declining because pasturage and soil were exhausted. The Commissioners were probably not correct to see nature as the major reason for the fall in the white population. It was more likely due to the extrusion of white tenants, a widespread process in South Africa accelerated by the Depression and the fact that labour was more difficult to extract from whites. In many sheep districts, the population as a whole actually increased in the first half of this century.

A similar argument in respect of sheep has recently been reiterated in a sophisticated form by Dean and Macdonald (1994). They compare stocking rates by district since the agricultural censuses began in 1919 to show an overall reduction, especially marked in the drier districts. By contrast, some grassland districts on the peripheries of the Karoo have experienced growing stock numbers. They cite the negative correlation between increased water provision and numbers as evidence that veld quality is the major constraint. They also note the tendency for farmers to switch from wool-bearing merino types to the hardier Dorper mutton sheep as evidence of overall decline in pastures. Dean and Macdonald examine and reject various other causes of lower stocking rates and 'conclude that the current livestock stocking rate in the semi-arid and arid rangelands of the Cape Province is unrelated to market forces or state policy but is determined by utilizable primary productivity of rangelands' (p. 281).

The implication of their argument is that falling numbers and reduced stocking rates, which I have cited as a possible factor in explaining improvements or stabilisation in veld quality, reflect an overall decline. Their findings sit uneasily with the botanical evidence of Hoffman and Cowling (1990). Roux's response to the Hoffman and Cowling article, not published, is also sceptical. He suggests that the thickening of vegetation which has occurred in some locations tends to be of less palatable species so that there is little evidence of increased carrying capacity (interview, July 1994).

Dean and Macdonald's conclusions, however, are by no means the last word. A number of points require far more systematic historical research. On the one hand, it would have to be shown that the massive investments into fencing and fodder have been at best neutral in their effects and at worst pointless. On the other hand, both state policies over the past half-century and farmers' strategies would have to be discounted as significant reasons for reduction in stock numbers.

It is unlikely that the second point could be sustained in a systematic historical analysis. Controlled marketing, price stabilisation, and subsidies on wool over many decades have almost certainly been a factor in containing sheep numbers. There is no doubt that the huge increase between 1928 and 1931 was largely a response to the collapse in wool prices during the Depression. Farmers tried to produce more in order to maintain their incomes and pay their debts. Since then, sheep numbers have generally diminished, rather than increased, when wool prices have fallen. Subsidised destocking programmes, particularly in the late 1960s, have also made some impact. While the fall in stock numbers in the mid-1960s was partly due to drought, that at the end of the decade was more likely the result of successful government reduction programmes. The number of small stock owned by commercial farmers has remained relatively stable since.

The switch to Dorpers (a cross between Black Persians and Dorsets) and mixed wool–mutton breeds such as Dohne merinos is a complex phenomenon which in the past few years is certainly linked to low wool and high mutton prices rather than just veld conditions. Factors such as labour costs (for shearing) also affect farmers' choices. In some districts where sufficient water is available, there has been a switch from woolled sheep to beef cattle for similar reasons (Beinart 1994). Fluctuations in the numbers of small stock kept largely for mutton earlier in this century are also unlikely to have been simply a reflection of deteriorating pastures. For example, the percentage of non-woolled sheep, including the hardy Black Persians, declined in the first few decades of this century at a time when pastures almost certainly were degrading. According to Dean and Macdonald's argument, they should have been increasing.

Lastly, it seems essential that the debate on reasons for the decline in numbers and on the state of the veld also take into account the views of those who often know it best at a local level: farmers and farmworkers. Over a century of debate on degradation, and over half a century of government propaganda following the 1932 Soil Erosion Act, have made some impact. Dean and Macdonald are aware of this possibility but do not give it great weight. On the basis of a limited number of interviews in the Eastern Cape and Free State, I would argue that farmers' strategies require more investigation. Many of the big sheep farmers have inherited their farms and come from wealthier and more educated backgrounds than earlier generations. At least some sheep farmers are deeply aware of environmental issues and they regard with some dismay the stocking practices of their fathers and grandfathers. They tend to be deliberately cautious about stocking levels – a caution which they see as having paid off in the serious droughts of the early 1990s. Most interviewees are of the opinion that the veld has stabilised or improved in their districts within their period of memory.

The interview material is not sufficient yet for hard conclusions to be drawn, and clearly caution is required in dealing with the perceptions of farmers and agricultural officers. Aside from the difficulties in evaluating oral evidence about the general condition of veld in a district, some farmers are clearly aware that political transformations in South Africa may place their rights to the land under question. In arguing that they are now good stewards, they are also asserting rights to land and more generally their belief in private property and the technical ideas that govern their pattern of land use. There are countervailing pointers in the evidence, such as degradation resulting from possible decline in underground water in some districts.

At the very least, however, interviews suggest that the views and practices of individual farmers matter and that pasture history must take into account developments on particular farms rather than just generalised arguments about districts and regions. On a number of occasions, I was shown neighbouring farms where the condition of the veld, separated only by a fence, was significantly different. There certainly are devastated farms, and even areas, but there is also evidence of improvement.

CONCLUSIONS

The arguments on stock numbers and degradation require more extensive debate and research. Botanical work is already rich and is accumulating rapidly (for recent summaries see Dean and Macdonald 1994; Hoffman *et al.* 1995; Milton and Hoffman 1994). By contrast, environmental history research which systematically builds in perspectives from political economy as well as ecology has hardly begun. Nevertheless, there are some general points that can be made.

First, it is important to begin to disentangle apartheid, and the white farmers who largely supported it, from the environmental condition of the farms. Apartheid has certainly contributed enormously to environmental degradation in homeland areas (although even here the ecological picture in some zones is less bleak than it is sometimes painted). But it is unclear whether the cosy relationship between organised (i.e. white) agriculture and the apartheid state over many decades has facilitated or undermined conservationist policies.

Second, there is a well-established discourse about long-term environmental degradation in South Africa, which is enormously persuasive in its general outlines. Some of it has been absorbed, perhaps uncritically, by the new radical environmentalists and green lobbies in the country. A good deal of it may be accurate but it is also an attractive instrument with which to condemn the past. Planning for the future, however, requires a more accurate understanding of changing environmental conditions.

Third, a significant number of farmers – whatever their ideological views – have also been influenced by conservationist thought. Although the farmers interviewed thus far have tended to be more progressive, it is quite clear that most were well informed about the environmental issues and planned their activities, such as stocking levels, partly with environmental safety in mind. They were not simply victims of nature.

Lastly, and most significantly for this collection, it is not very helpful to measure environmental change in terms of degradation of a pristine environment and call the result decay. Human survival requires environmental disturbance. Moreover, all the areas under discussion have been used by game and domesticated animals for a very long period and their ecology has not been static. There is no possibility of restoration short of the abandonment of stock farming completely and – if consistency is to be maintained – most other farming in most other areas. The evidence from the drier parts of South Africa, as well as many other parts of the country, while by no means clear, suggests that change has not simply been linear degradation. While decay is quite possible, a concept of transformation is often more useful.

Viewed over the longer term, fluctuations are possible. In order to understand these, it is essential that the long-term changes in the political economy of any agrarian region, including the effects of state intervention, are analysed. But it is also important that the complexities of interaction in the natural world are not simply read off from changes in political economy. Not only do particular local ecologies deeply shape agrarian forms, but they can certainly influence the patterns of transformation and/or degradation.[3]

NOTES

1 *Editors' note*: This oft-quoted number comes from a very shaky reworking of Acocks's figures by B. H. Downing (Environmental consequences of agricultural expansion in South Africa since 1850. *South African Journal of Science 74*: 420–2 (1978)). Downing actually gives the figure as 2.4km; Wilson seems to have misquoted him.

2 Cowling's position seems to be somewhat contradictory – though perhaps not entirely as he could argue that the veld has deteriorated in the long term, and that it is in poor condition, but that grass cover is now improving.

3 Donald Worster (1990) makes these points elegantly. The east coast of New England is often cited: two centuries of farming was undermined by the expansion of Mid-West and Plains agriculture, to be displaced eventually by forest regrowth and suburbanisation.

REFERENCES AND FURTHER READING

Acocks, J. P. H. 1975. *Veld Types of South Africa: Memoirs of the Botanical Survey of South Africa no. 40*, 2nd edn. Pretoria. (Part of a series published by the Botanical Research Institute, Department of Agricultural Technical Services, Republic of South Africa.)

Beinart, W. 1984. Soil erosion, conservationism and ideas about development: a Southern African exploration, 1900–1960. *Journal of Southern African Studies* 11(1): 52–83.

Beinart, W. 1993. The night of the jackal: sheep, pastures and predators in South Africa, 1900–1930. Unpublished paper – preliminary version published in *Revue Française d'Histoire d'Outre-mer* 80(298): 105–29.

Beinart, W. 1994. Farmers' strategies and land reform in the Orange Free State. *Review of African Political Economy* 21(61): 389–402.

Cape of Good Hope. 1877. *Report of the Commission upon Diseases in Cattle and Sheep* (G. 3–1877).

Cowling, R. 1991. Options for rural land use in Southern Africa: an ecological perspective. In M. de Klerk (ed.) *A Harvest of Discontent: The Land Question in South Africa*. Cape Town: Institute for a Democratic Alternative for South Africa, 11–24.

Dean, W. R. J., and Macdonald, I. A. W. 1994. Historical changes in stocking rates of domestic livestock as a measure of semi-arid and arid rangeland degradation in the Cape Province, South Africa. *Journal of Arid Environments* 26: 281–98.

Durning, A. 1990. *Apartheid's Environmental Toll*. Washington, DC: Worldwatch Institute.

Glacken, C. 1967. *Traces on the Rhodian Shore: Nature and Culture in Western Thought from Ancient Times to the End of the Eighteenth Century*. Berkeley: University of California Press.

Grove, R. 1987. Early themes in African conservation: the Cape in the nineteenth century. In Anderson, D. and Grove, R. (eds) *Conservation in Africa*. Cambridge: Cambridge University Press.

Grove, R. 1994. *Green Imperialism: Colonial Expansion, Tropical Island Edens and the Origins of Environmentalism 1600–1860*. Cambridge: Cambridge University Press.

Hall, T. D. 1934. South African pastures: retrospective and prospective. *South African Journal of Science* 31: 59–97.

Hall, T. D. 1942. Our veld: a major national problem. Pamphlet, Johannesburg.

Hoffman, M. T. 1993. The potential value of historical ecology to environmental monitoring. In Marais, C., and Richardson, D. M. (eds) *Monitoring Requirements for Fynbos Management*. Cape Town: Foundation for Research Development Programme Report Series no. 11, 69–86.

Hoffman, M. T., and Cowling, R. M. 1990. Vegetation change in the semi-arid eastern Karoo over the last 200 years: an expanding Karoo – fact or fiction? *South African Journal of Science* 86 (July–October): 286–94.

Hoffman, M. T., Bond, W. J., and Stock, W. D. 1995. Desertification of the eastern Karoo, South Africa: conflicting palaeoecological, historical and soil isotopic evidence. Forthcoming in *Environmental Monitoring and Assessment*. Foundation for Research Development Programme Report Series no. 37.

Huntley, B., Siegfried, R., and Sunter, C. 1989. *South African Environments into the 21st Century*. Cape Town: Tafelberg.

Milton, S. J., and Hoffman, M. T. 1994. The application of state-and-transition models to rangeland research and management in arid succulent and semi-arid grassy Karoo, South Africa. *African Journal of Range and Forestry Science* 11(1): 18–26.

Ramphele, M. (ed.). 1991. *Restoring the Land: Environment and Change in Post-apartheid South Africa*. London: Panos.

Roux, P. W. and Vorster, M. 1983. Vegetation change in the Karoo. *Proceedings of the Grassland Society of Southern Africa* 18: 25–9.

Roux, P. W., Vorster, M., Zeeman, P. J. L., and Wentzel, D. 1981. Stock production in the Karoo region. *Proceedings of the Grassland Society of Southern Africa* 16: 29–35.

Scoones, I. 1994. Politics, polemics and pastures: range management science and policy in Southern Africa. Paper presented to 'Escaping Orthodoxy: Environmental Change Assessments in Africa', Institute of Development Studies, University of Sussex, September 1994.

Sears, P. B. 1935. *Deserts on the March*. Norman: University of Oklahoma Press.

Shaw, J. 1873. On some of the changes going on in the South African vegetation through the introduction of the merino sheep. *Report of the British Association for the Advancement of Science, 43rd Meeting, Transactions of the Sections*, p. 105.

Taylor, C. C. 1935. *Agriculture in Southern Africa*. US Department of Agriculture, Technical Bulletin no. 466.

Tidmarsh, C. E. 1952. Veld management in the Karoo. Reprint no. 4, Grootfontein College of Agriculture (Pretoria: Government Printer), from *Farming in South Africa* 27(310): 4.

Union of South Africa, Senate. 1914. *Report of the Select Committee on Droughts, Rainfall and Soil Erosion* (SC 3).

Union of South Africa. 1923. *Final Report of the Drought Investigation Commission* (UG 49).

Union of South Africa. 1932. *Report of the Native Economic Commission, 1930–32* (UG 22).

Union of South Africa. 1951. *Report of the Desert Encroachment Committee* (UG 59).

Wilson, F., and Ramphele, M. 1989. *Uprooting Poverty: The South African Challenge*. Cape Town: David Philip.

Worster, D. 1979. *Dust Bowl: The Southern Plains in the 1930s*. New York: Oxford University Press.

Worster, D. 1985. *Rivers of Empire: Water, Aridity and the Growth of the American West*. New York: Oxford University Press.

Worster, D. 1990. Seeing beyond culture. In 'A roundtable: environmental history'. *Journal of American History* 76: 1078–106.

9

REFRAMING FOREST HISTORY

A radical reappraisal of the roles of people and climate in West African vegetation change[1]

James Fairhead and Melissa Leach

Editors' note In this chapter James Fairhead and Melissa Leach challenge some of the commonly held beliefs about forest change in West Africa. Like William Beinart (Chapter 8), they show the dangers of considering past change only in terms of destruction and show how the influence of people can produce what the Western-based environmental movement would perceive as a 'good' environment (in this case a more woody landscape). Their study highlights a divergence between research on people–vegetation relationships, which tends to be based on short (decade) time-scales of analysis, and research on climate–vegetation relationships, which takes longer (millennial) time-scales of analysis. They take a different approach and advocate analyses that examine the dynamic and complex interrelationship between people, climate, and vegetation; in order to do this they suggest that a century time-scale perspective is most appropriate (see also Jean Grove, Chapter 3). There are some interesting parallels between this analysis and the discussion in our conclusion (Chapter 12) comparing the conceptualisation of time in models of the physical world and models of society.

INTRODUCTION

West African vegetation history is generally understood as one of decline: of desertification, savannisation of forest, degradation of dry and humid forests, soil erosion, and biodiversity loss. For forests, the latest FAO assessment of forest cover change between 1980 and 1990 (FAO 1993) seemingly provides authoritative evidence for this decline, encapsulating in statistics a level of supposed recent change emotive enough to draw funding to international and national forestry and environment organisations. Concern is accentuated when forest loss at the decadal time-scale is presented as an acceleration of deforestation over preceding decades and centuries, wrought by increasing farming populations and forest exploitation.

The time frames used to describe forest cover change are integral to assessments of cause and responsibility, and are thus politically loaded. For instance, the supposition that deforestation is recent, and accelerating over a time-scale of decades, suggests that responsibility lies with present inhabitants and their social and economic conditions, making it imperative for today's environment and development institutions to act. Such imperatives are considerably reduced for deforestation occurring in earlier centuries, which seems remote from 'today's problems'. And over millennia, not only do forest cover changes fall into the hazy domain of prehistory, but the balance of responsibility can more easily tip from anthropogenic issues to natural factors such as climatic change.

This chapter uses both historical data and sources sensitive to inhabitants' own perspectives to gain precision concerning the extent and time-scale of forest cover change. We suggest that the extent of deforestation in West Africa this century is being hugely exaggerated in scientific and policy literature. This not only stigmatises local populations unnecessarily and supports inappropriate policies which further impoverish the poor, but also distorts analysis of the nature, causes, and consequences of vegetation change. Indeed, as we shall suggest, a reliance on apparently 'hard' data over the decade time-scale, extrapolated backwards over centuries, has obscured important instances of forest cover increase. These increases have anthropogenic causes which operate over longer time-scales than have previously been appreciated, as well as climatic causes operating over time-scales of centuries rather than millennia.

The need to evaluate assessments of anthropogenic vegetation change in West Africa more critically was stimulated by our earlier research which documented the social dynamics of forest cover change in the Republic of Guinea (Fairhead and Leach 1995a, b, 1996a, b; Leach and Fairhead 1994a, b, 1995). Originally framed in terms of deforestation at all these time-scales, the study came to reveal a striking contrast between the assessments of vegetation change driving policy, and historical data. Indeed, during the past forty years, which today's policy analysts consider to have been the most destructive in the forest–savanna transition zone, forest areas had increased at the expense of savanna. Historical and social anthropological research showed that the same evidence that scientists and policymakers had been taking to indicate deforestation was properly interpreted as indicating anthropogenically assisted forest regeneration. In particular, the research revealed how forest islands found in savanna owed their existence to inhabitants who had encouraged them to form around savanna settlements, rather than being relics of past forests. It showed how forest fallow (farmbush) had been established in grassy savannas, and was not, therefore, degraded forest as had been thought. It showed how palms had been established in savannas, rather than being relics of the savannisation of forest. In short, the research showed how a discourse of deforestation had both obscured ways in which inhabitants were successfully managing their landscape, and unjustly supported the imposition of draconian and inappropriate environmental policies.

Here we bring the critical perspective developed in Guinea to an analysis of the wider West African literature. We begin by detailing assertions made today concerning forest loss over the decade and century time-scale for Sierra Leone, Liberia, Côte d'Ivoire, Ghana, Togo, and Benin. We show how evidence provided for forest cover change rests on particular ways in which forest has been defined (with much forest lost or gained in the translation between different definitions); on key assumptions concerning the extent of 'original' forest cover; on selective interpretation of historical sources; and on the ways present vegetation forms are taken to suggest processes of vegetation change. We go on to provide historical data which question the evidence and assumptions made concerning 'original' and turn-of-the-century forest cover, and show how an appreciation of inhabitants' land-use practices allows for 'indicators of recent forest loss' to be interpreted quite differently.

This re-evaluation suggests that forest loss in West Africa during the present century may be only about 15 per cent of the level usually suggested. Moreover, exaggeration of losses has obscured the expansion of the forest zone on its northern margins, observable over a century time-scale, and the impact of both people and climate on this process.

ASSERTIONS OF RECENT FOREST LOSS IN WEST AFRICA

In recent years there has been a profusion of analyses of deforestation, dealing both with West Africa in general and with particular countries. Table 9.1 presents the recent FAO figures concerning the nature, extent and rate of

Table 9.1 Forest cover change, 1980–90, in West African countries

Country	Forest area (tropical rain forest and moist deciduous forest) (000s ha)	Forest loss/yr 1981–90 (000s ha)	% of total forest lost/yr
Benin	4,183	56.7	1.4
Côte d'Ivoire	10,831	119.4	1.1
Ghana	9,151	134.0	1.5
Guinea	6,565	86.6	1.3
Liberia	4,634	25.4	0.54
Sierra Leone	1,889	12.3	0.65
Togo	1,318	21.8	1.7

Source: FAO 1993

Table 9.2 Anthropogenic deforestation during the present century

Country	Forest area c. 1900	Present forest area (1985)	% loss this century
Benin	1,120,000	47,000	96
Côte d'Ivoire	14,500,000	3,993,000	72
Ghana	9,871,000	1,718,000	83
Liberia	6,475,000	2,000,000	69
Sierra Leone	not given	–	–
Togo	not given	–	–

Source: Gornitz and NASA 1985

forest loss over the decade time-scale (1980–90), showing the relentless demise of remaining forest area with more than 10 per cent lost over the decade. Table 9.2 presents the data for anthropogenic deforestation in West Africa during this century as suggested in the most comprehensive assessment made of it to date (Gornitz and NASA 1985), showing countries to have lost between 69 and 96 per cent of the forest area which they had at the turn of the century. Table 9.3 presents the most recent statement concerning the extent of present forest in relation to the 'original forest' (IUCN 1992), showing that with the exception of Liberia less than 13 per cent of the original forest cover remains.

Several analyses suggest that most 'original' forest has been lost during the present century, and that much of this loss is attributable to recent decades. Indeed if, as many authors do, one defines 'original forest' to be the area of today's forest zone – thus excluding the savanna regions – then most modern authors equate 1900 forest cover with 'original cover', assuming that the whole zone was then covered with unbroken forest. This is the case for Côte d'Ivoire (e.g. Gornitz and NASA 1985; Myers 1994; Parren and de Graaf 1995; Fair 1992; IUCN 1992; Gillis 1988), Liberia (e.g. Dorm-Adzobu 1985; Gornitz and NASA 1985; Parren and de Graaf 1995), Ghana (Gornitz and

Table 9.3 Present forest cover in relation to 'original' forest cover

Country	Original forest cover	Present forest area	Present forest as % of original
Benin	1,680,000	42,400	2.5
Côte d'Ivoire	22,940,000	2,746,400	12.0
Ghana	14,500,000	1,584,200	10.9
Guinea	18,580,000	765,500	4.1
Liberia	9,600,000	4,123,800	43.0
Sierra Leone	7,170,000	506,400	7.1
Togo	1,800,000	136,000	7.6
TOTAL	76,270,000	9,904,700	13.0

Nasa 1985), Sierra Leone (Myers 1980), and Benin (Gornitz and NASA 1985). Van Rompaey sums up the orthodox view that 'Only about 8 million ha of West African forest remained in the mid-eighties. This is some 20% of the precolonial area' (van Rompaey 1993).

Many of these authors suggest that the bulk of deforestation has occurred within the past fifty years. For Sierra Leone, for example, Myers wrote that

> as much as 5,000,000 ha may still have featured little disturbed forests as recently as the end of World War II. It is a measure of the pervasive impact of human activities that the amount of primary moist forest now believed to remain is officially stated to be no more than 290,000 ha.
>
> (Myers 1980, p. 164)

Several other modern authors do, however, consider 'original forest' to have been cleared earlier. Thus most analysts of Sierra Leone date deforestation not to the twentieth century but to the mid-nineteenth century, when Sierra Leone was a major exporter of timber (e.g. Dorward and Payne 1975; Millington 1985; Cole 1968). Equally, few authors suggest that Togo has lost large tracts of forest this century, instead dating their loss to pre-colonial periods in uses especially linked to iron extraction (Goucher 1981). And several authors consider that farming populations in Liberia and Ghana have been increasing since the sixteenth, not the twentieth century, dating deforestation accordingly (Hasselmann 1986).

RE-EXAMINING BASELINES FROM HISTORICAL SOURCES

Estimations of the extent of forest loss during this century (slightly earlier for Sierra Leone and Togo), or since 'original forest' was intact, need baseline information. In some cases, such baselines are derived from analyses of forest cover made early in the century, commonly drawing on the global forest assessment of Zon and Sparhawk (1923) and the sources which they used for West Africa such as Breschin (1902). In other cases, especially either where estimations are of 'original' cover or where original cover and turn-of-the-century cover are equated, baselines are deduced from climatic data, by delimiting a zone where forest (climatically) could exist, and assuming that at 'origin' (or in the immediately pre-colonial period) it did indeed exist in this zone.

More careful scrutiny of historical sources concerning forest cover at the turn of the century show that baseline estimates based uncritically on the work of early analysts such as Zon and Sparhawk are seriously misleading. Several early authors (including Zon and Sparhawk) themselves equated 'forest cover' with estimations made then of the extent of the forest zone (i.e. the zone of dense vegetation south of the savannas). Analysts today misrepresent these estimates as referring to intact forest cover. Other early sources show, instead, that at the turn of the century much of the forest zone was covered not by high forest, but by cultivated land, farmbush (land under rotational bush fallow), and savanna inliers. Furthermore, several large areas earlier defined as 'forest' were in fact palm forest. Taking these other sources into account provides a very different 'baseline' with which to compare present-day forest cover, as a brief country-by-country survey shows.

For Côte d'Ivoire, Zon and Sparhawk's assessment states: 'The dense tropical forest starts at the coast, and almost without break covers more than 12,000,000 ha. Beyond it there is wooded brush land which gradually merges into the Sahara Desert' (Zon and Sparhawk 1923, p. 868). In 1902, Breschin estimated from the accounts of early colonial travellers that the high forest zone covered about 15,000,000 ha. But these oft-cited calculations were in fact never accepted by the early French colonial forestry administration. Following a forest assessment mission in 1924, the forester Meniaud was explicit in his critique. Under the heading 'Statistical errors concerning the area of *grande forêt*' he suggested that while 'the areas given generally in statistics as being occupied by high forest are calculated according to the extreme limits ... [that is] 11 million hectares for the high forest of Côte d'Ivoire', this failed to take into account large 'empty spaces', or areas of farm or savanna land within these limits (Meniaud 1930). He recalculated the area of 'the primary forest, or that which is exploited only by the export timber industry' at only 8,000,000 ha (ibid.). In this, his figures are supported by the famous early botanist Chevalier, who spent two years examining Ivorian forests:

> The forest of Côte d'Ivoire [is] ... in reality less vast than one had originally thought. We think that it measures about 12,000,000 ha.... We believe that it is not exaggerated to consider half of the area of the supposed virgin forest as occupied by a forest of recent formation, much less rich in wood. The real virgin forest thus only covers 6,000,000 ha of Côte d'Ivoire, of 12,000,000 which it appears to occupy.
>
> (Chevalier 1909, p. 45)

The 'forest of recent formation' was either forest maintained within fallow cycles (ten to fifteen years' growth), or forest which had recently been abandoned from management.

In 1957, around the time when most modern authors start the stopwatch of most rapid destruction, the equally renowned French forester Aubréville asked, 'Where is the beautiful forest of Côte d'Ivoire?' He noted then that 'except in the almost deserted [southwest] region between the Cavally and Sassandra where clearance has not happened, the reserves of Eaux et Forêts (4,500,000 ha)[2] represent almost the only intact places which today permit the analysis of dense forest' (Aubréville 1957/58). When he was first in Côte d'Ivoire in the 1930s, he left no doubt that the vast majority of the country's forest was even then farm bush, except in a few uninhabited areas:

> In Côte d'Ivoire, the secondary forest covers a considerable extent which it is impossible to calculate. Along routes, one sees only such farmbush, *the few clumps of primary forest remaining* still disappearing year on year ... one can no longer hold any illusion. Entire regions are covered only in secondary forest.
>
> (Aubréville 1938, p. 239; emphasis added)

When the 'baseline' is reconstructed from these data, the rate of change in Ivorian forest cover looks very different. The orthodox estimates of forest loss suggest that Côte d'Ivoire had around 14.5-16.0 million ha of forest in 1900, began to lose it dramatically around 1955, and by 1990 had only about 3 million ha remaining, giving record-breaking estimates of forest loss for the period 1950-90 of around 325,000 ha/yr (cf. Bertrand 1983; Myers 1994; IUCN 1992). Yet if Côte d'Ivoire had only 6 million ha of forest at the turn of the century, even assuming loss to have begun only in 1950, the annual rate of loss declines to 75,000 ha/yr – less than a quarter of the supposed rate.

For Ghana, a similar critique can be applied to statistics concerning forest loss. Gornitz's estimate of deforestation during the present century is based on Zon and Sparhawk's figure of 9,871,000 ha in 1920. But Zon and Sparhawk's own breakdown of this figure shows it to include savanna forest (2,592,000 ha), which must be subtracted, leaving 7,279,000 ha, and is ambiguous about whether this corrected figure includes areas which were inhabited and farmed. That of the total forest (including savanna forest), only 3,628,000 ha were considered to be 'merchantable' for timber suggests that much of the rest might have been rotational bush fallow land. And this seems to be confirmed in the report of Meniaud's 1924 assessment (Meniaud 1930), which again explicitly criticised earlier analysts of the forest zone for failing to distinguish between areas within the forest zone, and areas which actually then carried forest. Drawing on official statistics, Meniaud noted the area of *Grande forêt* (humid forest) as 4,500,000 ha, but said that in reality this figure

> must be reduced, as in the high forest there are patches of high grasses and low bush of savanna type (of which the natural cause is not always evident); considerable areas have been completely brushed or opened by axe and fire for yam, maize, manioc, and often oil palms and kola trees are almost the only trees left standing.
>
> (Meniaud 1930, p. 537)

In particular, he noted that the forest of the whole eastern region was very 'cut into', and estimated that 'one can only count on 2,500,000 ha as veritable intact primary *grande forêt*, or exploited only by loggers'. This picture is qualitatively confirmed in Thompson's earlier survey of the forests of Ghana (Thompson 1910). Thus for Ghana, again, the turn of the century 'baseline' from which to calculate twentieth-century forest loss thus appears very much lower, and with it the extent of deforestation this century: not 92,000 ha/yr since 1900 (comparing Gornitz's 1900 assertion with present cover), but *c.* 10,000 ha/yr – only 11 per cent of the supposed rate.

For Benin, modern analysts who assert that the country has experienced extensive recent deforestation also draw on Breschin, and Zon and Sparhawk. In 1902, Breschin estimated the forest area to be *c.* 1,200,000 ha and Zon and Sparhawk suggest that 'after the first mile or two of sandy waste, the 50 mile wide coastal plain is for the most part covered with dense tropical forest' (1923, p. 861). Once again such figures are dramatically contradicted by other historical sources. An 1893 map detailing vegetation in coastal Dahomey clearly distinguishes dense 'forest' vegetation from open farmbush or grassland (at 1:100,000), showing large areas of the latter. This is confirmed by the accounts of early colonial botanists who travelled to the region, and stated clearly that even parts of so-called 'forest' described by the explorers whom Breschin and others drew on (e.g. Albeca 1894–5) were palm grove, albeit derelict in large parts. The botanist Chevalier, who spent six months examining the vegetation of Dahomey in 1909, suggests that:

> Oil palm is the principal wealth of the colony. The densest stands are found *between the coast and the southern border of the wooded and marshy region of Lama* [i.e. the major block of forest marked on the 1893 map], on a band 60km wide and 110km long, going from the frontier of Togo to that of Lagos; it is no exaggeration to say that within these limits, the palm covers all the land and gives almost everywhere its aspect to the landscape. In some places, the uncultivated bush seems to dominate, but when one examines it closely, one finds that this bush covers land in fallow; among the trees and high bushes live etiolated *Elaeis* [oil palm] in close rows, obscured by the wild vegetation.
>
> (Chevalier 1912, p. 30; emphasis added)

The early explorers had mistaken this palm-rich fallow for 'majestic forest'. The only dense 'forest' which Chevalier saw lay in very small patches in some swamps and on sacred sites, such as those in the eastern section of the so-called 'forest' zone which he considered to be relicts: 'from Sakete to Pobe one crosses swamps not used for crops, and covered with a high forest canopy in which one finds most of the beautiful species of Côte d'Ivoire. These islands are, however, of very small extent' (Chevalier 1912, p. 34).

For Liberia, the dramatic estimates of forest loss given by Gornitz, Parren and de Graaf, and others can also be re-evaluated, in part using historical data. The only early evaluations of Liberian forest cover did not differentiate between the area of the forest zone and the area of the forest. Zon and Sparhawk estimated 6,475,000 ha of forest, and Parren and de Graaf suggest 7,300,000 ha. These are, however, estimates of a climatically estimated 'forest zone', which was assumed then to have been unfarmed – an assumption which cannot be upheld given the documented significance of agricultural populations in the nineteenth century (cf. Massing 1985; Anderson 1870, 1874/1912; Seymour 1860). More detailed assessments from the 1950s (e.g. Mayer 1951; Haden-Guest *et al.* 1956) which suggest that Liberia had about 5,500,000 ha of high forest are probably better guides to what existed in 1900, given the stagnation of rural populations in the first half of the twentieth century. Not only has the extent of forest cover in 1900 been exaggerated, but, as Hasselmann (1986) argues, present cover has been underestimated. His recalculations using Republic of Liberia air surveys suggested that FAO estimates of *c.* 2,000,000 ha of forest remaining (FAO 1981) were less than half the actual areas. In short, the extent of conversion of high forest to bush fallow in Liberia during this century may represent between 10 and 20 per cent of the area suggested by most authors concerned with deforestation.

As in Liberia, a number of modern analyses use baseline estimates of 'original' forest cover derived not from uncritical use of past sources, but from the assumption that where forest climatically could exist today it did once exist. In general, these climatic estimates are far higher than those given by early observers, since they incorporate not only the whole of the 'zone of dense vegetation' as early observers described, but also areas considered to be derived savanna, i.e. parts of the Guinea savanna and forest–savanna transition zone to the north. Frequently, such areas – thought climatically able to support forest – are assumed to have been forest in the recent past, but to have been savannised through inhabitants' farming and fire-setting. Where these areas were savanna at the turn of the century, observers then did not include them within the forest zone. This differential treatment can account for large apparent increases in the area defined as 'original forest' between those analyses drawing on early sources, and those drawing on climatic assumptions. Thus for example while Chevalier considered the Ivorian forest zone to be 12 million ha in area, recent estimates of 'original forest' based on climatic assumptions and taking in supposed derived savanna suggest it to be 16 million (Myers 1994) or even an outrageous 23 million (IUCN 1992).

Thus even to the extent that it is useful to think in terms of an 'original forest cover' for West Africa – and we will question this below – it is clear from historical data that this was far from 'intact' as high forest at the turn of the century. If modern analysts are searching for periods of major forest loss, they may have to extend their analyses back to a different time-scale.

ONE-WAY DECLINE? RE-EXAMINING INDICATORS FROM THE PERSPECTIVE OF INHABITANTS' PRACTICES

Orthodox conclusions concerning forest loss are frequently supported by observations of present-day vegetation which supposedly indicate the existence of extensive forest cover in the recent past. This type of analysis – of process from form – has been long elaborated in the scientific literature (see Fairhead and Leach 1996b). Indeed, apart from oral testimony, it was the only form of evidence concerning vegetation history available to early botanists, who did not yet have access to time-series air photographs and documentary descriptions. Many of these early botanists were as convinced as today's analysts of recent and ongoing deforestation. However, many of the vegetation forms which they took to indicate the past existence of forest – such as 'relict' forest islands, 'relict' forest trees, and the presence of oil palms in savanna (the examples we consider here) – can be interpreted very differently from a perspective which takes into account inhabitants' land-use practices.

In Benin, Chevalier noted what he took to be relict forests and relict trees in 1909, and deduced from their existence that today's gap in the forest zone between Togo and Nigeria once supported a thin band of forest. Subsequently Aubréville deduced that forest patches and the presence of isolated forest trees were the residual of an earlier high forest cover:

> As testimony [to this forest] ... there remain the fetish woods, and some forests on the way to rapid destruction on the frontier with Nigeria, near Pobe ... a few isolated trees in the fields, sometimes a clump of ancient trees, a sacred grove, in the proximity of a village.
>
> (Aubréville 1937, p. 43)

More recent botanists and foresters such as Mondjannagni who have studied Benin's forests have concentrated their efforts on forest islands, treating them as 'dense, humid, semi-deciduous rainforest, veritable relicts of an ancient continuous forest cover' (Mondjannagni 1969).

Such deductions are the norm for the region. In his *Vegetation of Sierra Leone*, for example, Cole asserts that forest outliers in savanna regions probably originated as remnant forests; islands of original vegetation in 'a sea of grass or derived savanna' (1968, p. 81). Nyerges's analysis of the guinea savannas of northern Sierra Leone also takes forest outliers there as indicators of forest retreat, asserting that

> the history of the southern guinea savanna, in fact, is one of constant chipping away at the forest edge. ... The zones of 'derived' savanna or forest–savanna mosaic frequently marked on vegetation maps reflect this process of savannisation, in which disturbed forest sites are invaded by grasses that subsequently burn and prevent or retard the regrowth

of forest and the redevelopment of soil. Characteristics of this zone, which imply a history of degradation, include a sharp forest–savanna boundary, the presence of forest outliers and emergents in savanna, and the mosaic pattern of primary forest, secondary forest, farmland, and tall grass savanna that constitutes the transition between the forest and savanna zones.

(Nyerges 1987, pp. 327–8)

There are several reasons why these deductions concerning relict forests are invalid. First, many of the forest islands in Benin and Sierra Leone (as well as Côte d'Ivoire, Ghana, and Togo) are associated with settlements. In Guinea, we have shown how 'relict forest outliers' in the savannas of Kissidougou prefecture are not relicts at all, but are gradually established and enriched around villages which were initially sited in savanna. Such forests may thus be the product of settlement and active vegetation management (and tree planting) by inhabitants (Fairhead and Leach 1996a, b). Second, other sacred and 'relict' forest sites frequently cover the ruins of ancient villages and graves; sites with very specific and often super-fertile 'anthro-pogenic soils' which may support a dense vegetation where prior to habitation they would not (cf. Keay 1947; Thomas 1942; Sobey 1978). Third, sites to be rendered sacred may themselves be managed in very particular ways, including the planting of particular trees, and the sowing of particular termite species which would favour particular tree species development (cf. Iroko 1982; de Surgey 1988, 1994, cf. Fairhead and Leach 1994a).

That relict forests are anthropogenic might be gauged from their botanical composition. It is common in West Africa for tree species central to both ritual and everyday life to be planted or transplanted by inhabitants. Thus Benin's forest islands (as described by Chevalier 1912, 1910; Aubréville 1937; and Mondjannagni 1969) commonly have very few of the tree species associated with moist, semi-deciduous forest, and a preponderance among these of species known to be transplanted. Certain species are known for their importance in sacred performance and what is colloquially known as 'fetishism'. *Antiaris africana* (false Iroko) is planted both to install sacred altars in the region (cf. also de Surgey 1988, pp. 320–34), and for its bark, which can be cut and beaten into a traditional cloth. *Milicia excelsa* (Iroko/ Odu) is also a fetish tree, planted in sacred sites, in front of houses, and as a 'palaver tree' (cf. a o de Surgey 1988, pp. 320–34). *Triplochiton scleroxylon* is also a fetish tree *Ceiba pentandra* is not only a tree used for sacred purposes as throughout West Africa, but also grows rapidly from its cuttings, used very commonly both in fencing and fortification. In Guinea, this tree is planted (and sometimes trained with lateral branches or apical meristem being cut to shape the tree's growth) to initiate the further development of forest patches in savanna landscape. *Cynometra megalophylla* is known to be transplanted from riparian forest (Mondjannagni 1969). *Ficus* species are often planted both for fruit and for social purposes (Isert 1788; Gayibor 1986, p. 245). Other trees planted around sacred grove altars used by all Benin's people include *Newbouldia levis*, *Spondias mombin*, and *Dracaena arborea* (Mondjannagni 1969). Several of these trees are also used as staking poles for yam cultivation. In neighbouring parts of Nigeria *Khaya grandifolio* and *Canarium schweinfurthii* are noted as likely to have been planted (Lamb 1942). *Allanblackia floribunda*, *Carapa procera*, and *Pentaclethra macrophylla* used to be transplanted for their useful oil grains, and *Blighia sapida* and *Chrysophyllum africanum* are commonly planted for their fruit, as is *Daniella oblonga* (Chevalier 1910, 1912; Dalziel 1937).

That forest islands in Benin might have been the product of people's vegetation management, and enriched with socially valued tree species, was in fact noted by early travellers to the region. The most notable is Isert, who writes in a letter of 28 March 1785:

> The area around Fida [Ouida, southwest Benin] is one of the most attractive of all the places where the Europeans have settled in Guinea. The ground is level and blessed with meadows in which there are fresh water sources scattered all around. The farther I come into the Bight of Benin, the more enthusiastic I find the people in their worship of idols. At Orsu [Osu], around Christiansburg [now in Ghana], they have no public fetish temples; here they have more than thirty. I have seen some here which have several forecourts and a number of rooms, and are surrounded by beautiful trees. I like to go to such places because I always find those trees there which are rare in the country and are planted [*places à dessein*, in Gayibor's 1989 French translation] because of their rarity.
>
> (Isert 1788/1992, pp. 104–5)

Benin's 'forest islands' have a very similar flora to the particular 'anthropogenic' forests which came to cover ruined towns in old Oyo in Nigeria, owing their existence to habitation (Keay 1947), and to the *Kurmi* forest islands described in Nigeria by Lamb (1942), thought to have been established around villages because of habitation:

> *Kurmis* [forest islands] due to the protection by man are nearly always small and are usually the result of fire protection of the land round a village that is situated at the side of a stream bearing fringing forest.... But for fire protection, these would have been a closed savanna woodland type, for which both soil and climate are suitable. Fire protection, however, has allowed high forest to become established. This type may be seen at Gwada and Tagbare, north of Minna, where the climate could support high forest but soil conditions are not good enough till altered by man's interference.
>
> (Lamb 1942, p. 188)

Indeed, it is becoming increasingly apparent that the establishment of forest outliers around inhabited or ruined villages in the forest–savanna transition zone is generic and germane throughout West Africa. In Sierra Leone, Migeod noted this clearly in the 1920s:

> It is no uncommon thing to see a small forest round towns and villages, when there in none surviving anywhere else. Chief among the trees is the kola.... The Bombaces rear themselves above all the others covering much ground with their enormous buttresses. The origin of thick timber growth round a town was defensive purposes. The old stockades have taken root, and one may trace the lines of them in the big trees at the present day.
>
> (Migeod 1926, p. 334)

Apart from our own work in the Kissi and Kuranko speaking regions of Guinea, several recorded oral accounts speak of forest island establishment. For example, this emerges from oral history in Toma-speaking areas of Guinea:

> When the big silk-cotton tree [*Ceiba pentandra*] of Kuankan was planted, Jaka Kaman, when he founded the village, told all of the compound heads to dig up all of the young cotton trees. "When you dig up the cotton trees, go and plant them somewhere else. When you reach the [village site] boundary, start planting the trees back where you started." After that happened, everyone went and uprooted the cotton trees. Blessings and prayers were made before they were planted. Then the women went and watered them. That is why the cotton trees grew up. Three hundred and thirty three cotton trees were in Kuankan.
>
> (Wata Mamadi Kamara; edited from the manuscript notes of Tim Geysbeek *et al.*)

Forest island outliers are also found in more northern savannas – as a characteristic of Benin's landscape, for example in Savalou and Djougou; and from Bassila to Sokodé in Togo – and they all conceal villages (Aubréville 1937). These have generally been considered as 'natural', in this case by Aubréville as the conserved relicts of a past wetter climate, but a reconsideration is necessary, as it is for the peri-village forests found on the northern margins of the Ivorian forest zone.

Several authors who consider forest islands as relicts also consider the isolated 'forest' trees found on open land along roadsides and in fields as relicts. This is especially the case when they appear in their so-called 'forest form', with a tall, straight bole, supposedly indicating that they have grown up surrounded by dense forest. Yet certain trees characteristic of the transition zone can grow tall and straight spontaneously, even when in the open, especially when their lateral branches are lopped for fuelwood or other purposes; such trees include *Milicia excelsa*, *Antiaris africana*, and *Ceiba*

181

pentandra. It is quite possible, then, that the trees which authors such as Aubréville or Chevalier referred to as 'isolated relicts' grew into this 'forest form' in the open, preserved as they often are by farmers, rather than in a forest of which there is no other sign.

A critique of vegetation analysis in the Upper Guinean region should not stop at reinterpreting forest islands and relict trees. It also puts into doubt the validity of treating secondary thicket always as degraded forest. Susu farmers in northwest Sierra Leone, for instance, show a strong awareness of techniques to upgrade savannas to secondary forest thicket, or farmbush, fallows using a combination of intensive grazing and organic matter incorporation, stressing the role of termite activity in this (Leach and Fairhead 1994a).[3] Such upgrading of savannas into secondary forest thicket has also been noted in the Kuranko Loma mountain region (cf. Pocknell and Annalay 1995).

A similar upgrading of savannas has been noted in Togo's forest–savanna mosaic, where farming has classically been carried out in forest fallow fields, but where recent demographic pressures have resulted in the expansion of shifting cultivation into savanna areas. In some cases tree growth is encouraged in these fallows, and forest crops are then established. Successional stages in the establishment of forest depend on disturbance or 'accidental intervention by farmers', who, by initial cultivation, encourage the germination of pioneer, light-demanding forest species (e.g. *Harungana madagascariensis*), and then protect the developing forest vegetation (Guelly *et al.* 1993).

Many analysts of West African vegetation also treat the presence of oil palm trees as indicating the past presence of dense forest (Allison 1962; Keay 1959). They assume first, that where palms can grow, so forest can grow, and second, that where forest can grow, forest did grow. Nevertheless, both Chevalier (1912) and Aubréville (1937, p. 57) argue that much of Benin's palm forest was actually established in savanna, and thus cannot indicate the presence of past forest. Chevalier, in particular, noted the tendency for farming to lead to an *extension* of oil palm formations into the savannas:

> The large extension which the inhabitants have given, since some time, to the cultivation of crops, especially maize, leads to the development of palm groves, for in clearing, the inhabitants conserve the palms and start to look after them from the day that the soil is first used. It is no exaggeration to affirm that in Dahomey, all fields of maize made from the savanna/bush and sometimes also, unfortunately, from the forest (which has thus, and little by little, disappeared almost completely) become subsequently a palm plantation. One can thus hope that in a few years, the vast territory of 600,000–700,000 ha defined above could become an immense plantation of *Elaeis*.
>
> (Chevalier 1912, p. 30)

Palm forests were also noted by early visitors to Togo. For example, the Basel missionary, Burgi, made a protracted tour around most of Ewe territory; it was reported that he 'found actual forest *only* in Ewe, round Davie [Davie] and Darave [Dalave], where the oil palms and silk cotton trees (*bombax*) are so thick that one can walk for hours in shadow' (Burgi 1888). Yet given the predominance of palm trees and silk cotton, both of which were commonly established in savanna by farmers investing in what was then the highly profitable palm oil economy, it would appear that this forest could have been anthropogenic and established in savanna. This would be analogous to the anthropogenic palm forests known to have established at this time near the Volta River, just across the border in Ghana. In 1865, a British parliamentary select committee was told that 'There is a large plain near the Volta which once had no trees on it about 100 years ago, and the whole of that plain is now palm forest ... planted by the natives' (Johnson 1964, p. 21).

Aubréville claimed to be able to determine (from species composition/ biodiversity) those palm groves that had been established in forest and those which had been established in savanna. It is certainly the case that very recent palm groves established in savanna can be detected, for example if they contain *Parkia biglobosa*, a savanna tree kept for its valuable product, or if they are attacked by certain diseases. Nevertheless, it is less certain that these methods would be capable of identifying what vegetation much older palm groves were established in, since in this region, closed canopy formations such as dense palm groves very rapidly become diverse and acquire flora associated with forest. Thus finding a palm grove which contains forest species and no savanna ones may be no indication that it was established in forest.

An analogous interpretative problem surrounds a rather odd 'forest formation' which covers much of Benin, consisting essentially of baobabs (*Adansonia digitata*) in secondary forest thicket. Aubréville (1937) was astonished to see baobabs in a closed forest formation, and suggested that this was a degraded form of a forest type specific to Benin, within which baobabs were integrated. His opinion concerning this was crystallised when he examined the sacred grove of Foncome (between Athième and Parahoue). Here he found baobabs in an island of equatorial forest; he deduced that in this region baobabs' natural habitat is equatorial forest, and that baobabs in thicket and baobabs in savanna represent progressive stages of degradation of this form. Such historical deductions were in keeping with his pejorative views of African land management and of ongoing savannisation.

But this formation may be better interpreted in an opposite way: that the forest thicket was established by farmers in what were earlier baobab-rich savannas. As described by Mondjannagni (1969, p. 142), savannas can be thus upgraded to form thicket using special weeding practices – weeding before the burn – which limits the destruction caused by savanna fire. The baobab savannas are thus transformed into baobab-wooded thicket. This (anthropogenic) thicket typically contains young, vigorous forest trees such as *Mallotus oppositifolius, Antiaris africana, Dracaena arborea, Dialium guineense,* and *Uvaria chama,* as well as savanna trees such *Fagara xanthoxyloides* and *Dichrostachys glomerata.* The presence of baobabs in a sacred forest should properly have alerted Aubréville to the fact that this sacred forest was, in fact, an old village site in which it is common for baobabs to be established, and the forest patch was thus growing over very particular soils. We have noted such baobabs in Guinean forest patches, now overshadowed by forest trees in abandoned anthropogenic forest islands.

Indicators which have been used to deduce forest loss can, therefore, when read from a perspective which incorporates inhabitants' opinions and practices, testify instead to inhabitants' capacity to enrich their landscape with desired vegetation. Our emphasis here has been to draw attention to the problems of these 'indicators' in the light of our Guinea work, which showed how wrong they could be. This may not resolve the issue in other countries in every case, but certainly shows the need for more research to see how valid this critique may be in different instances.

CHANGES IN THE FOREST–SAVANNA TRANSITION ZONE OVER THE CENTURY TIME-SCALE

The previous section has highlighted many examples where inhabitants have been enriching vegetation over a time-scale of a century or more, in many cases rendering previously less woody vegetation more woody. At the northern margins of the forest zone, this evidence combines with other historical data to suggest that forest has actually been advancing into savanna over this time frame. In other words, the forest–savanna transition zone may not be a zone of derived savanna, but one of forest advance linked to its peopling.

In Côte d'Ivoire, evidence for encroachment of the forest zone on savanna comes from several sources: from studies of vegetation change which use historical and present data; from oral history in today's savanna, transition, and forest zones; and from archaeological evidence. Several accounts suggest that areas now well within the forest zone have been savanna in the recent historical past. For example, in a forest area southeast of the Baoulé V, Ekanza heard from elders in the Moronou region that

> the Agni traditions of origin make reference to the savanna as being the form of vegetation which dominated Moronou at the period of settlement. Village sites during this period of invasion were chosen in function of the openness of savanna. [The village of] Brobo for which

the origin goes back to the earliest time of Agni settlement in Moronou, was built on a site in savanna.

(Ekanza 1981, p. 59)

In a taped interview in 1981, the 85-year-old Eonan Messou suggested that the important historical figure Nanan Sangban 'abandoned Brobo, where game was getting rare, for the preferential site of Bongouanou, which offered at once savanna, water and animals.... The foundation of Arrah and other villages obeyed the same imperatives.' From Bongouanou to Arrah, tradition reports, in referring to the period of settlement, a game-filled savanna extending as far as the eye could see (Ekanza 1981, pp. 59–60). Ekanza finds it difficult to credit the implication that forest expansion has been so rapid as to absorb all of the savanna in the space of only two centuries (from the eighteenth century when Agni arrived to 1907 when early French colonial observers saw only forest or forest fallow). Yet in neighbouring Guinea the encroachment of forest into farmed savanna in the transition zone has covered tens of kilometres in only three decades (Fairhead and Leach 1996b; Leach and Fairhead 1994b).

Again in Côte d'Ivoire, west of the Baoulé V, in Gouro country, and again in what is now the forest zone, several Gouro villages carry names which indicate the past existence of savanna; Koumodje, meaning forest island in the middle of savanna, or Deragon, meaning open savanna (Deluz 1970). Such a vegetation history is entirely consistent with ecological findings in Côte d'Ivoire since the 1960s. In the savannas of the Baoulé V, for example, villagers themselves suggest that 'where one cultivates, the forest advances', and research on forest dynamics at the forest–savanna boundary shows just that (Spichiger and Blanc-Pamard 1973). As Adjanohoun writes:

When one asks Baoulé elders about the origin of savannas, they affirm that their ancestors, 200 years ago, found the same vegetation formation which they call *kakie*, meaning wooded savanna. The old cultivators recognise that they have contributed to the degradation of islands of dense forest, but they make you observe that the fallow on their ex-forest fields, exhausted by cropping, regrows not as *kakie*, but in forest fallow. They affirm equally to remember the existence of *kakie*, once found within the dense forest, and which has today disappeared, entirely invaded by this [forest]. For them, the grassy savannas are natural, and there is not a current phenomenon of savannisation. On the contrary, it is the forest which gains on the savanna, and this despite their action.

(Adjanohoun 1964, p. 31)

A second study shows forest cover increasing over savanna between the 1950s and 1970s in the Beomi region (Spichiger and Lassailly 1981). These trends continue in the Baoulé V savannas: 'Lamto savannas are characterised

185

by ... their dynamic evolution towards forest' (Menaut and Cesar 1979, p. 1197). 'In 20 years, tree density in annually burnt plots has increased by c. 30%' (Dauget and Menaut, unpublished data cited in Menaut et al. 1991, p. 136). The latter suggest that this might be due to 'wave-like' or cyclical patterns over long phases – of which they are seeing an intensification – and that extreme events or episodic, concurrent disturbances should then be responsible for the maintenance of savannas in the very long term. But this overlooks the evidence above: that, very probably, in the areas east of Lamto the savannas have not been maintained.

In 1937, in the Wenchi-Kintampo district of central Ghana, the forester Vigne noted 'many cases of closed forest encroaching on savanna forest'. He attributed this to climatic rehumidification. Foggie (1957) also noted that: 'In the north-west, the savannah at one time extended much further south. The forest reserves north and west of Sunyani are rapidly changing from savanna woodland ... to closed forest'. He differed from Vigne in his explanation, suggesting that this could only be due to depopulation – but nevertheless shows it to be happening at a time of population increase (Foggie 1957, p. 132).

On the northern margins of the forest zone, then, where other ecological conditions are marginal for forest, it seems that human influence may be a crucial factor in enabling and encouraging the formation of forest in savanna where otherwise this would be unlikely. With the exception of forest islands, much of this forest extension into savanna is kept within the agricultural bush–fallow rotations which are its *raison d'être*. Yet it is possible, indeed probable, that several areas of high forest towards the northern margins of the forest zone represent such upgraded vegetation which has subsequently 'escaped' from bush–fallow rotation as a result of depopulation and/or forest reservation. This is the probable origin, for example, of high forest within the Ziama reserve of Guinea (Fairhead and Leach 1994b), for parts of the Loma Mountain reserve of Sierra Leone (Pocknell and Annalay 1995), of the Tain Tributaries forest reserve in Ghana (Foggie 1957), of the Togodo forest reserve in Togo (Cornevin 1969), and of the Okomu forest reserve in Nigeria (Jones 1956).

Farther south, such dynamics of depopulation and repopulation are associated with historical fluctuations between high forest and bush fallow over a time-scale of several centuries. In many cases, periods of depopulation in recent centuries can be traced to the slave trade and associated warfare, and later to wars surrounding colonisation. The timing of such depopulation allowed the high forest which regrew to be reserved by early colonial authorities. The presence of old village sites still identifiable by their distinctive vegetation in nearly all West Africa's forest reserves is testimony to their past occupation. Well-documented examples include the Gola forest reserve in Sierra Leone (Small 1953; Unwin 1909). In Ghana, it seems to apply to Numia forest reserve in the heart of Ashanti country and to

Mpameso in northwest Ashanti (Taylor 1960; cf. also examples in Thompson 1910). That this is general to Liberian forests is suggested by Voorhoeve (1965), to those of Côte d'Ivoire by Mangenot (1955), and to those of Togo by Cornevin (1969).

Clearly, then, West African forest cover change has not been a one-way process. In the heart of the forest zone, there have been periods and areas of conversion of high forest to farmbush and farmland, and periods when the opposite has happened. The emphasis on deforestation found in the policy and scientific literature has obscured attention to these non-linear dynamics, and to the forest advance into savanna which has occurred in the northern margins of the forest zone in recent centuries.

CLIMATIC CHANGE

While people's landscape enrichment practices and the dynamics of human populations have been important influences on changes in the extent of the forest zone over the past few centuries, this process also needs to be examined in relation to climate change. Those studies which note forest advance into the savannas have generally interpreted it in terms of climatic rehumidification. Indeed in most cases, such forest advance is seen to be despite human activities, which are still seen as having a negative impact on forest cover. Only a few studies argue – as we do here – that people's practices and increasing climatic humidity may be complementary forces in the transformation of savannas to forest or forest fallow.

Vigne, for example, attributed forest advance in Ghana largely to climatic change, considering that the Gold Coast was at the time experiencing a wet cycle, an indication of which being the rise in water level in Lake Bosumtwi (Vigne 1937). He was perhaps the first to argue that in tension zones, relatively small climatic changes may have important influences. He went on to argue that

> it is difficult to account for the large extent of 'savanna forest' in the area I studied by assuming a larger population in the past; I consider it is due partly to drier climatic conditions in the past, especially as measured in rainfall and humidity over the fairly short dry season.
>
> (Vigne 1937, pp. 93–4)

Vigne's analysis long pre-dates Aubréville's own (1962) retheorisation of forest–savanna dynamics in terms of climatic rehumidification. Aubréville, who in the 1930s and 1940s pioneered the idea of forest recession and savannisation under human management and desiccation (Aubréville 1949), which still dominates policy circles, came to change his mind over this issue. He eventually suggested that 'the climatic conditions which permitted the establishment of savannas were those of a relatively recent period, and ... we are seeing again today the development of forest colonisation following climatic rehumidification' (Aubréville 1962, p. 30).

In the analysis of forest–climate relationships in West Africa, most emphasis has been placed on a very long time-scale, with early authors suggesting that the soils and vegetation might be responding to the effects of general rehumidification since the extreme arid phase pre-12,000 BP (Aubréville 1962; Avenard *et al.* 1974). However, there is now growing evidence that rehumidification from more recent intervening dry phases may be more important, suggesting that climatic change over shorter time-scales may be having an appreciable influence on vegetation. Several recent studies have noted that there was a very dry phase in West African coastal climates around 3000 BP (in Gabon, cf. Schwartz 1992; in Cameroon, by Maley, pers. comm. in Schwartz 1992, also Talbot 1981 and pers. comm.). The importance of climatic change over this time-scale to the forest zone has been underscored by Schwartz:

> Palynological studies have shown that the climatic extension of open formations [c. 3000 BP] preceded the arrival of farmer-metallurgist populations usually considered to be very aggressive towards their environment; the present savannas are not of anthropic origin.
>
> (Schwartz 1992, p. 359)

That there have been even more recent – if less profound – dry phases, including a recent dry phase c. 1300–1850, is suggested by Nicholson (1979, 1980). Nicholson's historical evidence is supported by evidence from the levels of Lake Bosumtwi in Ghana, which were in decline, or low, from c. 700 years ago until they began to rise c. 200 years ago, forcing lake shore villages to be abandoned because of inundation (Talbot and Delibrias 1977). As Nicholson (1980) notes, descriptions and rainfall measurements in Freetown in the 1790s, for example, suggest that normal rainfall may then have been perhaps only half its present levels (cf. Winterbottom 1969).

Thus the scale of the temporal dimension to West Africa's forest–climate relationship may be properly measurable not in thousands, but in hundreds of years. It would also not be inconceivable for today's vegetation to be responding simultaneously to recovery from drier conditions at each of these time-scales – i.e. from protracted and deep aridity around 12,000, from a short deep aridity 3,500 and from a relatively arid phase 700–200 years ago. Whether expressed through effects on soil, soil fauna and flora, or vegetation distribution, lag effects from each dry phase might remain relevant, interfering with present responses to more recent climatic variation.

Furthermore, evidence of climate change undermines the basic methods and assumptions which have been used to assess forest loss. If climate is changing and appears to have undergone rehumidification, the assumption of a constant, climatically determined 'forest zone' cannot be upheld. The assumption that where forest could exist today, it did once exist, ceases to be valid, since parts of these zones may in the past have been too dry to support forest. Furthermore, the notion of an 'original' forest cover – however calculated – ceases to be useful in the light of constant climatic change. If 'baselines' from which to calculate subsequent forest cover change are to be used, then they need to be historically specified.

CONCLUSIONS

Many of the assertions made today concerning deforestation in West Africa during the present century seem to be wildly inaccurate. Table 9.4 sums up the potential exaggeration of conversion of dense humid and semi-deciduous forest to farmbush and savanna over the century time-scale. It suggests that deforestation may have been less than a fifth of that suggested in the international literature.

These exaggerations derive, in large part, from the use of particular methodologies for assessing vegetation change. As we have seen, these rarely make direct use of historical data (such as past remote sensing/air photographs, ground photography, archival descriptions, and oral evidence). Rather they deduce the nature and extent of past vegetation and the time-scale of its loss from observations of present vegetation and theories concerning its origin and dynamics; and from the presumption that everywhere that forest *can* exist today (under given climatic and soil conditions), it *did* exist in pristine form until it was either degraded or lost completely to savanna. Furthermore, these studies commonly presume that supposed forest loss was caused primarily by human habitation and land use, that the human action affecting forest cover is

Table 9.4 Forest loss during the present century: broad orthodoxy and revision estimates compared

Country	Orthodoxy			Revision		
	1900 cover	Present cover	Forest loss	1900 cover	Present cover	Forest loss
Ghana	9.9	1.6	**8.3**	2.5	1.6	**0.9**
Côte d'Ivoire	14.5	2.7	**11.8**	6	2.7	**3.3**
Benin	1.1	0.42	**0.68**	0.5	0.42	**0.08**
Togo	not available	0.13	–	not available	0.13	–
Sierra Leone	5*	0.5	**4.5**	0.09**	0.5	**gain: 0.41**
Liberia	6.5	2.0+	**4.5**	5.5	4.8++	**0.7**
TOTAL			**29.8**			**4.57**

Sources: Orthodox estimates based on Gornitz and NASA 1985 (past) and IUCN 1995 (present) unless otherwise indicated. Revision estimates based either on critical sources presented earlier in the chapter or on indicated references: *Myers 1980; **Unwin 1909; +FAO 1981; ++Hasselmann 1986.
Note: All figures in 1,000,000 ha

189

relatively recent, and that ancient populations and land use were either insignificant, or forest benign. In challenging the findings produced through this methodology, this chapter shows the importance of historical methods for examining changes in vegetation and land use. It emphasises the need to reframe human impact over a longer time-scale and climatic impact over a shorter one, and the need to see vegetation change in terms of pathways conditioned by unique interactions of these variables.

Orthodox methodologies and assumptions about time-scale have also obscured important evidence of forest advance encouraged by people, especially at the northern margins of the forest zone. It can be argued – reasonably we think – that the vegetation forms of forest island, palm groves, and baobab thicket have all been read backwards by those specialists who consider them either as degraded forms, or, in the case of forest islands, as relict islands of conservation in an otherwise decimated landscape. Historical data and sources sensitive to inhabitants' own perspectives suggest that they may, instead, be anthropogenically enriched forms. If forest islands are found to have been planted, or to have become established on old village sites, or both – that is, to be anthropogenic or on anthropogenic soils – then treating these forests as testimony to a now lost natural vegetation would be incorrect. If people have established dense palm groves in savanna, and if baobab bush is not degraded forest but enriched savanna, then the whole edifice of suppositions concerning 'West Africa's past vegetation' begins to look very weak. In fact, an entirely different set of propositions about vegetation change could be forwarded. If coastal West Africa has been experiencing a period of climatic rehumidification, are its forest islands early outposts of encroaching forest vegetation? And at periods in history, have people been encouraging the formation of forest vegetation forms in savanna, whether in their farm and fallow practices, their palm management, or their settlement strategies?

To the extent that these propositions hold (and detailed place-specific fieldwork is certainly needed to examine them), then there are far-reaching implications for research and policy. An edifice of analysis in many disciplines is built on the foundation idea of past and ongoing forest recession under habitation. This has been the backdrop as much for analysis of economic, social, and political history in the region and for cultural and ecological anthropology as it has been for studies of ecological dynamics and global environmental change. Equally, the figures purporting to show the nature and rate of forest loss support an academic industry which seeks to gain precision concerning its demographic, economic, and political causes (e.g. Pearce and Brown 1994; Barnes 1990; Allen and Barnes 1985), and its regional and global climatic consequences (e.g. Gornitz and NASA 1985; Ledant 1984–5; Huguet 1982; CAB International n.d.), as well as its medical and veterinary ones (e.g. Dorward and Payne 1975; Monnier 1980). Reframing West African forest history similarly demands a reframing of such analyses. Studies examining the relationship between forest change and demographic change are a case in point:

of course people can and do reduce forest cover, but they can also install it, or assist its installation. This observation undermines the model used by FAO's 1993 assessment, which assumed a negative relationship between people and forests (FAO 1993), and demands precision about not just how many people there are, but what they are doing, and what the prior vegetation was.

But the concerns here are not merely academic, for orthodox suppositions about the time-scales, causes, and effects of deforestation have been equally central to the justification and elaboration of often draconian environmental policies. If this chapter is correct, then the people who have been living in the c. 25 million ha of the forest zone which has been only mythically deforested during this century are owed an apology. They have been blamed for damage which they have not caused, and have paid heavily for this in policies aimed at controlling their so-called environmental 'vandalism', and at removing their control over resources in favour of national and international guardians (cf. Tanzidani 1993; Fairhead and Leach 1994a, 1995b, 1996b; Leach and Mearns forthcoming). Deforestation orthodoxies may, then, be adding unnecessarily to impoverishment, social upheaval, and conflict, while denying inhabitants their own history and views of vegetation change.

NOTES

1   Research for this chapter was made possible by a generous 'Global Environmental Change' Research Fellowship to J. F. from the Economic and Social Research Council (ESRC) of Great Britain. Its theoretical perspective is rooted in previous joint research in Guinea funded by ESCOR of the Overseas Development Administration. Responsibility for the conclusions drawn and any errors of fact or judgement nevertheless rests with the authors alone.

2   We have assumed that the original text has a misprint, meaning 4.5 million ha when it states 45 million.

3   We are grateful to Kate Longley for facilitating these conversations during her research in Kukuna, Sierra Leone.

REFERENCES

Adjanohoun, E. 1964. Végétation des savanes et des rochers découverts en Côte d'Ivoire centrale, *Mémoire ORSTOM no. 7*, Paris.

Albeca, A. 1894–5. L'avenir du Dahomey. *Annales de Géographie* 4: 166–221.

Allen, J. C., and Barnes, D. F. 1985. The causes of deforestation in developing countries. *Annals of the Association of American Geographers* 75(2): 163–84.

Allison, P. A. 1962. Historical inferences to be drawn from the effect of human settlement on the vegetation of Africa. *Journal of African History* 3(2): 241–9.

Anderson, B. 1870. *Narrative of a Journey to Musardu, Capital of the Western Mandingoes*. New York: S. W. Green.

Anderson, B. 1874/1912. *Narrative of the expedition despatched to Musahdu by the Liberian government under Benjamin K. Anderson, Senior, in 1874*, ed. Frederik Starr. Monrovia, Liberia: Monrovia College of West Africa Press.

Aubréville, A. 1937. Les forêts du Dahomey et du Togo. *Bulletin du Comité d'Étude Historique et Scientifique de l'Afrique Occidentale Française* 20: 1–148.

Aubréville, A. 1938. La forêt coloniale: les forêts de l'Afrique Occidentale Française. *Annales de l'Académie des Sciences Coloniales, IX.* Paris: Société d'Éditions Géographiques, Maritimes et Coloniales.

Aubréville, A. 1949. *Climats, forêts et désertification de l'Afrique tropicale.* Paris: Société d'Édition de Géographie Maritime et Coloniale.

Aubréville, A. 1957/58. A la recherche de la forêt en Côte d'Ivoire. *Bois et Forêts des Tropiques* 56: 17–32; 57: 12–28.

Aubréville, A. 1962. Savanisation tropicale et glaciation quaternaire. *Adansonia* 2(1): 16–84.

Avenard, J.-M., Bonvallot, J., Latham, M., Renard-Dugerdil, M., and Richard, J. 1974. *Aspects du contact forêt–savane dans le centre et l'ouest de la Côte d'Ivoire: étude descriptive.* ORSTOM: Abidjan.

Barnes, R. F. W. 1990. Deforestation trends in tropical Africa. *African Journal of Ecology* 28: 161–73.

Bertrand, A. 1983. La déforestation en Zone de Forêt en Côte d'Ivoire. *Bois et Forêts des Tropiques,* 202 (1983), 3–17.

Breschin, A. 1902. La forêt tropicale en Afrique, principalement dans les Colonies françaises. *La Géographie* 5: 431–50; 6: 27–39.

Bürgi, E. 1888. Reisen an der Togoküste und im Ewegebiet. *Petermanns Mitteilungen* 34 (8): 233–7.

CAB International (Development Services) (n.d.) *Deforestation in Africa: A Literature Review.*

Chevalier, A. 1909 Les massifs montagneux du nord-ouest de la Côte d'Ivoire. *La Géographie* 20: 207–24.

Chevalier, A. 1910 Le pays des Hollis et les régions voisines. *La Géographie* 21: 427–33.

Chevalier, A. 1912 *Rapport sur une mission scientifique dans l'Ouest africain (1908–1910),* 12 January. Paris: Missions Scientifiques.

Cole, N. H. A. 1968. *The Vegetation of Sierra Leone.* Njala, Sierra Leone: Njala University Press.

Cornevin, R. 1969. *Histoire du Togo.* Paris: Editions Berger-Levrault, 3rd edn.

Dalziel, J. M. 1937. *The Useful Plants of West Topical Africa.* London: Crown Agents.

de Surgey, A. 1994. *Nature et fonction des fétiches en Afrique Noire.* Paris: L'Harmattan.

Deluz, A. 1970. *Organisation sociale et tradition orale: les Gouro de Côte d'Ivoire.* Paris: Mouton.

Dorm-Adzobu, C. 1985. Forestry and forest industries in Liberia. An example in ecological destabilization. *Internat. Inst. für Umwelt und Gesellschaft pre-85,* 14. Berlin.

Dorward, D. C., and Payne, A. I. 1975. Deforestation, the decline of the horse and the spread of the tsetse fly and trypanosomiasis (Nagana) in nineteenth century Sierra Leone. *Journal of African History* 16(2): 241–56.

Ekanza, S.-P. 1981. Le Moronou à l'époque de l'administrateur Marchand: aspects physiques et économiques. *Annales de l'Université d'Abidjan, Série 1, Histoire* 9: 55–70.

Fair, D. 1992. Africa's rain forests – retreat and hold. *Africa Insight* 22(1): 23–28.

Fairhead, J., and Leach, M. 1994a. Termites, society and ecology: perspectives from Mande and Central West Atlantic regions. Paper presented to the African Studies Association UK Biennial Conference, University of Lancaster, 5–7 September.

Fairhead, J., and Leach, M. 1994b. Contested forests: modern conservation and historical land use in Guinea's Ziama reserve. *African Affairs* 93: 481–512.

Fairhead, J., and Leach, M. 1995a. False forest history, complicit social analysis: rethinking some West African deforestation narratives. *World Development* 23(6): 1023–36.

Fairhead, J., and Leach, M. 1995b. Reading forest history backwards: the interaction of policy and local land use in Guinea, 1893–1993. *Environment and History* 1(1): 55–92.

Fairhead, J., and Leach, M. 1996a. Enriching the landscape: social history and the management of transition ecology in the forest–savanna mosaic (Republic of Guinea). *Africa* 66: 1.

Fairhead, J., and Leach, M. 1996b. *Misreading the African Landscape: Society and Ecology in a Forest-Savanna Mosaic*. Cambridge: Cambridge University Press.

FAO 1981. Tropical Forest Resources Assessment Project, Rome. Forest Resources of Tropical Africa, Part I and II (Country Briefs).

FAO 1993. Forest Resources Assessment 1990. Tropical countries. *FAO Forestry Paper* 112.

Foggie, A. 1957. Forestry problems in the closed forest zone of Ghana. *Journal of the West African Science Association* 3: 141–7.

Gayibor, N. L. 1986. Écologie et histoire: les origines de la savane du Bénin, *Cahiers d'Études Africaines* 101102, XXV1-1-2, 13–14.

Geysbeek, T. 1994. A traditional history of the Konyan (15th–16th century): Vase Camara's epic of Musadu. *History in Africa* 21: 49–85.

Gillis, M. 1988. West Africa: resource management policies and the tropical forest. In Repetto, R., and Gillis, M. (eds) *Public Policies and the Misuse of Forest Resources*, Cambridge: Cambridge University Press.

Gornitz, V., and NASA 1985. A survey of anthropogenic vegetation changes in West Africa during the last century – climatic implications. *Climatic Change* 7: 285–325.

Goucher, C. L. 1981. Iron is iron 'til it is rust: trade and ecology in the decline of West African ironsmelting. *Journal of African History* 22: 179–89.

Guelly, K. A., Roussel, B. and Guyot, M. 1993 Initiation of forest succession in savanna fallows in SW Togo. *Bois et Forêts des Tropiques* 235: 37–48.

Haden-Guest, S., Wright, J. K. and Tecloff, E. M. (eds). 1956. *A World Geography of Forest Resources*. American Geographical Society Special Publication 33.

Hasselmann, K. H. 1986. Liberian forests, geoecological ponderabilities. *Liberia Forum* 2/3: 26–60.

Huguet, L. 1982. Que penser de la 'disparition' des forêts tropicales? *Bois et Forêts des Tropiques* 195: 7–22.

Iroko, A. F. 1982. Le role des termitières dans l'histoire des peuples de la République Populaire du Bénin des origines à nos jours. *Bulletin de l'IFAN* 44B(1–2): 50–75.

Isert, P. E. 1788 (1992). *Letters on West Africa and the Slave Trade. Paul Erdmann Isert's Journey to Guinea and the Caribbean Islands in Columbia (1788)*, translated from the German and edited by Selena Axelrod Winsnes. Oxford: Oxford University Press.

IUCN (International Union for Conservation of Nature). 1992. *The Conservation Atlas for Tropical Forests: Africa*, ed. Sayer, J. A., Harcourt, C. S., and Collins, N. M. IUCN.

Johnson, M., 1964. Migrant's progress, Part I. *Bulletin of the Ghana Geographical Association* 9(2): 1–27.

Jones, E. W. 1956. The plateau forest of the Okomu forest reserve. *Journal of Ecology* 53 and 54.

Keay, R. W. J. 1947. Notes on the vegetation of old Oyo reserve. *Farm and Forest*, Jan.–June, 36–47.

Keay, R. W. J. 1959. Derived savanna – derived from what? *Bulletin IFAN, Series A* 2: 427–38.

Lamb, A. F. 1942. The *kurmis* of Northern Nigeria. *Farm and Forest* 3: 187–92.

Leach, M., and Fairhead, J. 1994a. The forest islands of Kissidougou: social dynamics of environmental change in West Africa's forest–savanna mosaic. Report to ESCOR of the Overseas Development Administration, July.

Leach, M., and Fairhead, J. 1994b. Natural resources management: the reproduction of environmental misinformation in Guinea's forest–savanna transition zone. *IDS Bulletin* 25(2): 81–7.

Leach, M., and Fairhead, J. 1995. Ruined settlements and new gardens: gender and soil ripening among Kuranko farmers in the forest–savanna transition zone. *IDS Bulletin* 26(1): 24–32.

Leach, M., and Mearns, R. (eds) (forthcoming). *The Lie of the Land: Challenging Received Wisdom in African Environmental Change and Policy*. IAI series Issues in African Development. London: James Currey.

Ledant, J.-P. 1984–5. La réduction de biomasse végétale en Afrique de l'Ouest: aperçu général. *Annales de Gembloux* 90: 195–216; 91: 111–23.

Mangenot, G. 1955. Étude sur les forêts des plaines et plateaux de la Côte d'Ivoire. *Études Éburnéennes* 4: 5–61.

Massing, A. W. 1985. The Mane, the decline of Mali and Mandinka expansion towards the south Windward Coast. *Cahiers d'Études Africaines* 97(25, 1): 21–55.

Mayer, K. R. 1951. *Forest Resources of Liberia*. Information Bulletin 67 of the United States Department of Agriculture, Washington, DC.

Menaut, J. C. and Cesar, J. 1979. Structure and primary productivity of Lamto Savannas, Ivory Coast. *Ecology* 60(6): 1197–1210.

Menaut, J. C., Gignoux, C., Prado, C. and Clobert, J. 1991. Tree community dynamics in a humid savanna of the Côte d'Ivoire: modelling the effects of fire and competition with grass and neighbours. In Werner, P. (ed.) *Savanna Ecology and Management: Australasian Perspectives and International Comparisons*. Oxford: Blackwell Scientific Publications, 127–37.

Meniaud, J. C. 1922. *La forêt de la Côte d'Ivoire et son exploitation*. Préf. de M. le Gouverneur Antonetti. Introduction et considérations générales sur le pays et les habitants, de M. M. Larre. Paris: Publications Africaines.

Meniaud, J. 1930. L'arbre et la forêt en Afrique Noire. *Académie des Sciences Coloniales. Comptes Rendus Mensuels des Séances de l'Académie des Sciences Coloniales: Communications* 14: 1929–30.

Migeod, F. W. H. 1926. *A View of Sierra Leone*. London: Kegan Paul, Trench & Trubner.

Millington, A. C. 1985. Soil erosion and agricultural land use in Sierra Leone. Ph.D. thesis, University of Sussex.

Mondjannagni, A. 1969. *Contribution à l'étude des paysages végétaux du Bas-Dahomey. Annales de l'Université d'Abidjan*, série G, Tome 1, fasc. 2.

Monnier, Y. 1980. Meningite cérébro-spinale, harmattan et déforestation. *Cahiers d'Outre Mer* 130 (April–June): 103–22.

Myers, N. 1980. *Conversion of Tropical Moist Forests*. Washington, DC: National Academy of Sciences.

Myers, N. 1994. Tropical deforestation: rates and patterns. In Brown, K. and Pearce, D. W. (eds) *The Causes of Tropical Deforestation: The Economic and Statistical Analysis of Factors Giving Rise to the Loss of Tropical Forests*. London: UCL Press.

Nicholson, S. E. 1979. The methodology of historical climate reconstruction and its application to Africa. *Journal of African History* 20(1): 31–49.

Nicholson, S. E. 1980. Saharan climates in historic times. In Williams, M. A. J., and Faure, H. (eds) *The Sahara and the Nile: Quaternary Environments and Prehistoric Occupation in Northern Africa*. Rotterdam: A. A. Balkema, 173–200.

Nyerges, A. E. 1987. The development potential of the Guinea savanna: social and ecological constraints in the West African 'Middle Belt'. In Little, P. D., and

Horowitz, M. M. (eds) *Lands at Risk in the Third World: Local Level Perspectives.* Boulder, CO, and London: Westview.

Parren, M. P. E., and de Graaf, N. R. 1995. The quest for natural forest management in Ghana, Côte d'Ivoire and Liberia.

Pearce, D. W., and Brown, K. 1994. Saving the world's tropical forests. In Brown, K., and Pearce, D. W. (eds) *The Causes of Tropical Deforestation: The Economic and Staistical Analysis of Factors Giving Rise to the Loss of Tropical Forests.* London: UCL Press.

Pocknell, S., and Annalay, D. 1995. Report of the University of East Anglia Expedition to the Loma Mountains.

Schwartz, D. 1992. Assèchement climatique vers 3000 B.P. et expansion Bantu en Afrique centrale atlantique: quelques réflexions. *Bull. Soc. Géol. France* 163(3): 353–61.

Seymour, G. L. 1860. The journal of the journey of George L. Seymour to the interior of Liberia: 1858. *New York Colonization Journal*, 105, 108, 109, 111, 112.

Small, D. 1953. Some ecological and vegetational studies in the Gola Forest Reserve, Sierra Leone. M.Sc. dissertation, Queen's University, Belfast.

Sobey, D. G. 1978. Anogeissus groves on abandoned village sites in the Mole National Park, Ghana. *Biotropica* 10 (2): 87–99.

Spichiger, R., and Blanc-Pamard, C. 1973. Recherches sur le contact forêt–savane en Côte d'Ivoire: étude du recru forestier sur des parcelles cultivées en lisière d'un îlot forestier dans le sud du pays baoulé. *Candollea* 28: 21–37.

Spichiger, R., and Lassailly, V. 1981. Recherches sur le contact forêt–savane en Côte d'Ivoire: note sur l'évolution de la végétation dans la region de Béoumi (Côte d'Ivoire centrale). *Candollea* 36: 145–53.

Surgey, A. de. 1988. *Le système réligieux des Evhe.* Paris: L'Harmattan.

Talbot, M. R. 1981. Holocene changes in tropical wind intensity and rainfall: evidence from southeast Ghana. *Quaternary Research* 16: 201–20.

Talbot, M. R. and Delibrias, G. 1977. Holocene variations in the level of Lake Bosumtwi, Ghana. *Nature* 268: 722–4.

Tanzidani, T. K. T. 1993. Les problèmes sociaux dans les reserves de faune du Togo. *Cahiers d'Outre Mer* 46(181): 61–73.

Taylor, C. J. 1960. *Synecology and Silviculture in Ghana.* Edinburgh: Thomas Nelson, and University College of Ghana.

Thomas, A. S. 1942. A note on the distribution of *Chlorophora excelsa* in Uganda. *Empire Forestry Journal* 21: 42–3.

Thompson, H. 1910. *Gold Coast: Report on Forests.* Colonial Reports – Miscellaneous no. 66. London: HMSO.

Unwin, A. H. 1909. *Report on the Forests and Forestry Problems in Sierra Leone.* London: Waterlow and Sons.

van Rompaey, R. S. A. R. 1993. Forest gradients in West Africa: a spatial gradient analysis. Doctoral thesis, Department of Forestry, Wageningen Agricultural University, The Netherlands.

Vigne, C. 1937. Letter to editor. *Empire Forestry Journal* 16: 93–4.

Voorhoeve, A. G. 1965. Liberian high forest trees. A systematic botanical study of the 75 most important or frequent high forest trees, with reference to numerous related species. Ph.D. thesis, Agricultural University of Wageningen.

Winterbottom, T. M. 1969. *An Account of the Native Africans in the Neighbourhood of Sierra Leone.* London: Cass.

Zon, R., and Sparhawk, W. N. 1923. *Forest Resources of the World.* New York: McGraw-Hill.

10

ECONOMIC ACTION AND THE ENVIRONMENT

Problems of time and predictability

Malte Faber and John L. R. Proops

Editors' note This chapter is unlike most of the others, in that although it is looking at how present human actors interact with their future, its concern is not so much prediction as an attempt to construct the outlines of a conceptual theory of how we can or should address an unpredictable future. Its specific concern is with the inadequacy of economics to address the issues which environmental change raises. It sees economics as trapped in a Newtonian physics-based model – which in a sense therefore dismisses time, in that fundamental parameters are assumed not to change while equilibrium is attained. It therefore proposes a model for economics based more on evolutionary biology – in which it is recognised that there are fundamental and unpredictable evolutionary changes in the technological base of society, which human actors recognise, and which limit their scope for action. The authors' conclusion is that the existing system of economic liberalism cannot be expected to cope with environmental challenges – a point which Max Wallis also touched upon. Although the authors do not come up with a new system to replace the old, they do systematically address some of the main issues to be faced.

INTRODUCTION

In this chapter we consider the relationships between economic actions and the environment, particularly with regard to time and predictability. We know that economic actions may be environmentally harmful, as attested by present concerns regarding air and water pollution, ozone layer depletion, and global warming. On the other hand, economic action can be environmentally beneficial, or at least mitigate negative effects. Examples here include the fitting of catalytic converters to cars, the move from coal to oil and gas as the principal fossil fuels (reducing carbon dioxide emissions), and improvements in the standards of the thermal insulation of buildings.

We shall argue that the environmental outcomes of economic actions are often difficult to predict. We shall further argue that this difficulty of prediction is a necessary consequence of the temporal nature of the modern economic system of the Western world. In particular, we shall seek to show that policymaking regarding environmental issues must take account of five factors.

1   Economic activity is goal-oriented, or teleological (from the Greek *telos* – goal/aim/end). By their nature, teleological activities must take place in time, with a certain minimum period needing to elapse between the intention for the action being formulated and the action being completed. The time that must elapse in this way we call the '*telos* completion time'.

2   Economic actions are made by agents who are often relatively 'short-sighted' concerning the wider consequences of their actions, particularly with respect to effects that occur at later dates. The temporal concept we use to analyse this phenomenon we term the 'time horizon'. For illustrative purposes we shall often employ the term 'entrepreneurial time horizon'.

3   Most economic activity involves capital goods, i.e. goods whose purpose is to help satisfy human desires in the future rather than the present. Such capital goods are limited in their useful lives; the period of usefulness of a capital good we term 'the lifetime of capital'.

4   Many economic actions involve the introduction of novelty, which puts limits on the time period over which predictions can be made. The time period over which we can make predictions concerning a particular economic activity is defined as the 'foreseeable future'.

5   Economic actions can have long-term consequences for the natural environment; the time period over which an economic action affects the natural environment we call the 'total impacts lifetime' of that action.

We shall try to show that the interaction between these five time concepts gives rise to fundamental, and probably insuperable, difficulties concerning the formulation of complete and intertemporally coherent plans for managing economy–environment interactions over time. However, our message is not meant to be entirely bleak, as we consider that identifying the nature of this inherent problem of prediction will enable the formulation of policy which avoids the failures of 'optimal' planning. Instead, the way is open for the formulation of environmental policy which recognises the limits to prediction, by being open to revision, and precautionary in nature.

TELOS, TELEOLOGICAL SEQUENCE AND *TELOS* COMPLETION TIME

We begin our discussion with an analysis of teleology as, we shall argue, it is the existence of goals, and goal-oriented behaviour, which holds the key to understanding the nature of economy–environment interactions, and the limits of prediction in this area.

Telos

We observe that humans have goals; the Greek expression for a goal is a *telos* (pl. *tele*). Thus we speak of behaviour that is oriented towards a goal as teleological. It is clear that economic behaviour, involving choice between alternative outcomes, is teleological.

It is important to realize that teleological explanations can be, and often are, used in all branches of science, not just social science (Faber *et al.* 1995a). For example, the path taken by a light ray through a heterogeneous medium can be understood using Hamilton's approach, which shows that the path taken is that which minimises the time of travel of a light ray. This formulation can be interpreted as being teleological, as it explains the path of the ray in terms of an outcome (i.e. shortest travel time) which has yet to occur when the light ray begins its travel.[1] In biology, the behaviour of organisms can be seen as goal oriented (e.g. seeking out food), although such behaviour can also be seen as deterministic responses to existing conditions.[2]

Teleological sequence

To pursue a *telos*, it is necessary to pursue a strategy which has a structure in time. Here it is useful to speak of different temporal stages. If the time order of the stages is fixed, or has only a limited range of variation, then we will speak of a teleological sequence (Faber and Proops 1993, p. 70). To give an example, if one wants to make iron (i.e. this is the *telos*), one needs first to mine the iron ore, coal, and limestone. Then in a second stage the coal must be turned into coke. Finally, the coke, ore, and limestone can be combined in the blast furnace, and smelting commenced. This teleological sequence consists of three stages, with an invariable temporal order. A biological example is breeding by bees. First, the beehive has to be built; then the wax cells have to be produced; next the cells must be filled with nutrients; only then can the eggs be laid. Again, this teleological sequence has an invariable time structure, in this case involving four stages.

Telos completion time

Any particular teleological sequence will take a certain time to complete; that is, to achieve the *telos*. The time that must elapse to achieve a certain *telos* may be somewhat elastic; for example, one might wait between stages of the operation (teleological sequence), or by more work make some of the stages rather shorter. However, the fact that the sequence must be performed in a definite order, and that each stage takes a certain time, means that there is a certain minimum time that must elapse in carrying out a teleological sequence. This minimum time we call the 'telos completion time'.[3]

Of course, goals may build on goals. For example, the telos may be the manufacturing of steel, but there is as yet no furnace. If the minimum time to produce the furnace is T_f, and the minimum time to produce a tonne of steel in the completed furnace is T_s, then the telos completion time T_{fs} for making steel is given by $T_{fs} = T_f + T_{fs}$.

The telos and the ignored

One way in which taking a teleological approach to decision making is revealing is that it allows the distinction between the telos, which is actively being pursued, and other effects which may result from this striving towards the telos. As these other effects are often excluded from consideration by the agent, we refer to them as the 'ignored'.

For example, a business person may have as a telos the achievement of a certain status within their firm, with the consequent bad, but ignored, effects that this single-mindedness may have on their marriage or health, for example. Similarly, in the West there has been an emphasis by governments on the achievement of economic growth. Although since the 1960s there have been voices raised against the pursuit of this telos, because of its various negative effects (e.g. Mishan 1969), until recently these negative aspects of growth were ignored.

In the neoclassical economics literature on the environment, the concept of the 'ignored' also appears, in a special form. This is the case of 'externalities', which are defined to be consequences of an action which have an effect on social welfare, but which are not mediated by markets, and which therefore do not have market prices (Tietenberg 1994, p. 37). However, our concept of the ignored is wider than that of externalities, for when a particular telos is being pursued, even market-mediated effects may be dismissed, or excluded from consideration.

Using the terminology we have developed elsewhere (Faber et al. 1992), we might say that very often the agent pursuing a telos is in a state of 'closed ignorance' concerning the ignored. That is, the agent wilfully refuses to admit the possibility of knowledge of any negative outcomes associated with the telos. We consider this state of closed ignorance to be characteristic of many economic agents (e.g. individuals, firms, governments), and to be one of the fundamental obstacles to rational decision-making.

TELEOLOGY AND ENTREPRENEURIAL TIME HORIZONS

Having described human activity as teleological, with associated teleological sequences and telos completion times, we now turn to the role of humans as decision-making agents. In this we seek to explore more closely the roots of teleology in human behaviour, in particular how *tele* are formulated and how teleological sequences are established.

The approach of most economists to decision-making stresses the rationality of agents. That is, economic agents will tend to take actions which are to their benefit, and avoid actions which are to their disbenefit. For example, a hungry person will tend to eat, rather than simply look at food. Similarly, those wanting to buy a good they cannot yet afford will tend to save, rather than spend their income contrary to their long-run desires. This normal 'economic' model of human behaviour has two main limitations, among others. First, it assumes individuals know their aims explicitly; second, it assumes that individuals do not have simultaneous, but contradictory, aims. There is ample evidence that neither assumption holds in many cases. For example, decision-making is a notoriously difficult exercise for many (perhaps most) people; we often feel we 'do not know our minds' concerning a choice between some outcomes. Also, many people find a conflict between two 'parts' of themselves. Duty may suggest one should get up and go to work; sloth demands one stay in bed. Hunger tells one to eat; a desire to become slimmer urges abstention.

Decision-making over time: discounting

The problem of decision-making over time is even more difficult. Most individuals seem to have a strong desire to satisfy wants immediately, while their 'better selves' tell them to forgo present consumption to allow future desires to be satisfied. This problem of giving greater weight to present than future needs is usually represented in the economics literature by the concept of 'discounting'. For example, if an individual has a discount rate of 10 per cent per annum, consumption today will be 1.1 times more desirable than consumption next year, and 1.21 times more desirable than consumption in two years' time. For our analysis we find it useful to supplement the notion of discounting[4] with a further concept relating human desires to actions over time. This is the notion of the 'time horizon'.

The entrepreneurial time horizon

We know, from casual experience, that human concerns are not indefinite in their temporal extent. Individuals and organisations seem to formulate views on the world that are of finite duration. For example, it is often supposed that the 'long run' for social concerns is up to the lifetime of grandchildren (i.e. fifty years); some governments have five-year plans; firms plan products over perhaps three to five years; some individuals consistently act as if 'there were no tomorrow'. In all these cases the economic agents are acting with finite time horizons, so that periods later than their time horizon effectively have an infinite discount factor. The use of the concept of a finite time horizon need not preclude the use of the concept of discounting. Rather, up to the end of the time horizon a normal (finite) discount factor would be used, while, as noted above, after the end of the time horizon there would be effectively infinite discounting; this implies that one does not care at all about what happens after the end of the time horizon.[5] It also, of course, implies a point of discontinuity.

However, not only do different individuals and organisations have different time horizons, but even one individual may in general have different time horizons, because all of us have different but simultaneous social roles. For economic gain we may have a short time perspective. Thus an owner of an ice-cream shop may, as a business person, be interested only in the sales of the present summer. At the same time she may have a longer view concerning her retirement and may, therefore, sign a long-run insurance contract. As a voter she may have an even longer view, voting for policies which protect national parks in perpetuity. For these reasons it is expedient to distinguish the different time horizons in these roles. In the context of the environment, it may often be useful to speak of the entrepreneurial time horizon, the consumer time horizon, or the public time horizon, although of course there are many other ways we could identify and classify differing human behavioural time horizons. For reasons of illustration we restrict our representation in the following mainly to the first of these.

Regarding environmental issues, we can immediately see that if social responses are limited by finite time horizons, then many long-term environmental issues may be ignored altogether. Deforestation is a good example. Since the time of Aristotle, in Western Europe successive generations have commented on the reduction of forest cover, and the consequent increasing scarcity of wood, for construction and fuel purposes. Even though every generation has known that there was a problem, and has had the means to alleviate the problem, through forest conservation and management, none has done so. The possible solutions to the problem were so long run they would come to fruition outside the time horizons of economic agents.

Telos completion time and the entrepreneurial time horizon

We now relate the concept of the entrepreneurial time horizon, discussed above, to teleology. We can now define an entrepreneurial time horizon in terms of teleological (i.e. goal-oriented) behaviour, as follows. The entrepreneurial time horizon for teleological behaviour is that time span over which the *telos* is pursued.

We have seen that to achieve a *telos* one is, in general, restricted to a corresponding teleological sequence, comprising several stages. Hence each teleological sequence has a certain time length, which we called the *telos* completion time. From this it follows that to achieve this *telos*, the entrepreneurial time horizon must be at least as long as or longer than the corresponding *telos* completion time. For example, a building firm may require one year to build a new house (i.e. its *telos* completion time is one year), while it has an entrepreneurial time horizon of five years. To give another illustration, the typical entrepreneurial time horizon for a newly elected Western politician, whose *telos* is re-election, is four years. Thus any programmes with major long-run social returns, but short-run social costs (such as environmental legislation), are unlikely to be pursued, as their *telos* completion time exceeds the entrepreneurial time horizon of the goal-oriented agent.

In summary, if a *telos* is to be pursued, it is necessary to have an entrepreneurial time horizon of the agent which is at least as long as the corresponding *telos* completion time. Thus we have a lower bound for the entrepreneurial time horizon related to a certain *telos*.

CAPITAL GOODS

Having outlined the concepts of teleology, *telos* completion time, and the entrepreneurial time horizon of economic agents, we now consider another economic concept, capital, which has had a major effect on long-run economy–environment interactions. Now it is evident that capital goods have finite useful lives. Any good which is put to some physical use must experience degradation, and eventually become impossible to use for its previously determined productive role.[6]

We can therefore define the 'lifetime of a capital good' as the period for which the capital good serves towards the *telos* for which it was produced. We can then relate the lifetime of capital to the entrepreneurial time horizon for a decision-maker, in the following way. If a decision-maker has a short entrepreneurial time horizon, then it is likely that the capital goods used to pursue the *telos* will have a short lifetime. This is because, in general, more 'effort' is required to construct capital goods with a longer lifetime than a shorter one. If one further assumes a moderate amount of rationality in the decision-maker, there will be a tendency to construct capital goods which will have been 'used up' by the time the entrepreneurial time horizon is reached. Therefore we might expect that the entrepreneurial time horizon of the decision-maker will define an upper limit on the lifetime of the capital goods used to pursue the *telos*.

However, we cannot establish a simple relationship between the *telos* completion time and the lifetime of capital produced towards the corresponding *telos*. This is because the achievement of the *telos* may require various types of capital goods, in sequence, so that capital goods with short lifetimes could be used to pursue a *telos* with a long *telos* completion time. For example, if the *telos* is building a new town, the *telos* completion time may be several decades. However, the capital goods used to build it will have a much shorter life span.

On the other hand, it is conceivable that the lifetime of capital goods may actually exceed the *telos* completion time. This is because the nature of the capital good required to achieve the *telos* may be such that it is inherently long-lasting. For example, some steam engines produced in the nineteenth century could still be used, even though they now have no useful purpose to serve. We know also of equipment in chemical plants which has been in use for more than five decades.

NOVELTY, PHENOTYPE, AND GENOTYPE

In this section we establish, at some length, various fundamental conceptual categories that will allow us to discuss the limits to prediction for economic systems, particularly with respect to environmental issues. We shall take an evolutionary perspective,[7] involving the use of the concepts of 'genotype' and 'phenotype'.

First, however, we begin with some remarks on the approaches of standard economics. It is generally accepted that, at least until the 1980s, historical time was neglected by the neoclassical approach, the main paradigm of economics (see, for example, Hicks 1973; Solow 1985, pp. 329–31). It is further well known that Newtonian physics had a strong influence on the shape of neoclassical economics (see, for example, Mirowski 1984). Of paramount importance in the neoclassical approach is the concept of an equilibrium. Of course, a particular equilibrium can only remain as long as no novelty occurs, for in such a case the equilibrium under consideration would cease to exist. Because of these difficulties new branches in economics, in particular evolutionary economics, were developed in order to cope with problems of novelty, especially with regard to technical progress (see, for example, Nelson and Winter 1982). For a different direction see the approach initiated by Arthur (1989), which is described well in a journalistic way by Waldrop (1992, pp. 31–52, 327–35). Since we have elsewhere dealt at length with problems of evolution, novelty, surprise, and ignorance, and discussed the relevant literature, we refer the reader to Faber and Proops (1993), Faber *et al.* (1992, 1994), and the literature cited therein.

Biological evolution

We generally suppose that any organism, whether animal or plant, belongs to a 'species'; the normal definition of a species is a group of organisms which can interbreed. The process of breeding creates new members of the same species, and this continuity is mediated by genetic material, with the genetic material of any organism being a blending of the genetic material from its parents.

The genetic material can be thought of as the 'blueprint' for the organism, and is known as the 'genotype' of the organism. Of course, exactly how the 'potentiality' of the organism, as given by the genotype, will be 'realised' depends also on the environmental conditions in which the organism finds itself. A poor diet will stunt the growth of a person, even if that person's genotype indicates the potential to become very tall. The realisation of the genotype, the actual form and capabilities of the organism, is known as the 'phenotype' of that organism. The way an organism (or species) reproduces itself is therefore:

Phenotype → Genotype → Phenotype → Genotype, etc.

The genotype determines (in part) the nature of the phenotype, while the genotype is transmitted to the phenotype of the next generation through the nature and activities of the current phenotype.

The evolution of phenotypes

One aspect of biological evolution is the evolution of phenotypes. Consider a population of various species of organism, and suppose that a new species is introduced into this population. For example, European rats were introduced into New Zealand by the nineteenth-century settlers. The populations of indigenous species are presumed to have been fairly constant in the preceding decades or centuries. However, with the introduction of a new species, a disturbance to the system is introduced. If the rat had proved to be rather a feeble competitor in New Zealand, it would have died out and the old equilibrium would have been re-established. In fact, it proved to be a fierce competitor, and soon wiped out many indigenous species. Clearly, in these circumstances the population will evolve towards a new phenotypic equilibrium, with rats as an important species in that new equilibrium.

The evolution of genotypes

Evolutionary theory in biology seeks to explain how species evolve. It proposes a two-part scheme: variation and natural selection.

In variation, the genotype of a single organism is modified. We now recognise that this is usually through 'mutation' of the genetic material, because of some outside influence such as cosmic rays. This variation in the original genotype will give rise to a modified phenotype for that organism's offspring. Natural selection can now come into play. This new phenotype will, very often, be less effective than the original (unvaried) phenotype. Consequently, the new phenotype will not be successful in breeding and so it will not survive in the population. However, very occasionally the new phenotype will be more effective than the old, so it will survive in the population.

We note that 'natural selection' is another name for the 'phenotypic evolution' discussed above. Thus we see that part of genotypic evolution is the process of phenotypic evolution.

Evolution and the generation of novelty

If we perturb a population of phenotypes (e.g. by a forest fire), we disturb the relative abundances of the species. However, if the system is sufficently 'resilient' (Holling 1973), over time the system will tend to evolve back to its original state. Nothing new happens, so we can predict the long-run outcome. We can say phenotypic evolution does not generate novelty, so it is a predictable process. (This concept of predictability we develop further in the next two sections.) On the other hand, genotypic evolution does generate novelty. This novelty is, we shall argue, unpredictable.

Evolution and economic activity

Just as biological systems can be described in terms of genotypes (potential-ities) and phenotypes (realisations), so can economies. The set of 'techniques of production' available to an economy, together with the preferences of consumers, the legal system in force and the natural resources available, specify the potentialities (i.e. genotype) of the economy. The economic activity that occurs, in terms of the quantities and prices of goods produced, the distribution of income and wealth, and the environmental impact of this activity, is the realisation of these potentialities (i.e. the phenotype).

Also, as in the case of biological systems, that part of economic activity which represents genotypic evolution (alteration of potentialities, or genera-tion of novelty), we shall argue, is unpredictable. On the other hand, economic activity corresponding to phenotypic evolution (realisations) may be predictable.

FORESEEABLE FUTURE AND PREDICTABILITY

In this section we now introduce a fundamental concept, that of the 'foreseeable future', which we define in relation to natural, economic and social 'processes'. We define a 'process' to be a means of transformation, whether natural and spontaneous, or as a reflection of human intervention and will. The extent to which the future is foreseeable will therefore depend on the extent to which such processes are predictable.

The predictability of processes

We first wish to distinguish between two types of forecasting of future events; we call these 'prediction' and 'speculation'.

Prediction we take to mean statements about future events which are based on a theory derived from experience. We find it useful to use this definition, even for cases where it later turns out that the theory is 'wrong'. For example, the Greek astronomers devised the 'Ptolemaic' system on the structure and working of what we now call the solar system. In their theory the earth is at the centre, with the sun, moon and planets orbiting about it. Further, they assumed the orbits to be composed of various combinations of circular movements ('epicycles'). The precise movements of the heavenly bodies were matched by their theory through the appropriate choice of combinations of epicycles. This theory allowed very accurate predictions of the movements of the planets; however, we now regard the theory as incorrect. Instead, we now theorise that the sun is at the centre of the solar system, and the orbits of the planets are nearly perfect ellipses, through the effects of universal gravitation.

By contrast, speculation involves statements about future events when there is no underlying theory based on experience, and indeed it may simply be 'plucked from the air'. However, the crucial distinction regarding speculation is that some theories supporting statements about the future are not based on experience. For example, various religious cults make rather frequent forecasts of the impending end of the world, usually based on the interpretation of religious writings. Clearly there is an implicit theory behind these forecasts, but this theory is based on almost no empirical evidence. Therefore such forecasting we term speculation rather than prediction.

The predictability of phenotypic evolution

Using our above definitions of prediction, we can now proceed to an analysis of what types of process are, in principle, predictable. We see that there are two conditions that must be met:

1   We must be able to establish a body of experience (empirical knowledge) over time concerning the process.
2   On the basis of that experience we must be able to formulate a theory.

It should be stressed that these conditions are for predictability in principle, not for the validity of any particular prediction. It is generally the case that predictions are to a lesser or greater extent incorrect, because of the limitations of data collection and theorising.

We noted above that phenotypic evolution involves the realisation of a given set of potentialities; that is, there is no emergence of novelty. Therefore a system which is evolving phenotypically allows observations on that system which permit the formulation of theories about the system's behaviour. So in principle a system undergoing only phenotypic evolution is predictable.

On the other hand, a system undergoing genotypic evolution also alters its potentialities, so observations of it should capture the emergence of novelty. As, by definition, novelty is something about the precise nature of which one cannot have a theory, genotypic evolution is unpredictable.

The foreseeable future of predictability

The above discussion on the predictability of phenotypic evolution, and the unpredictability of genotypic evolution, allows us to offer a concept of the 'foreseeable future' of systems. We define the foreseeable future of a system to be that length of time over which we can sensibly make predictions; that is, the length of time over which the system exhibits only very limited genotypic evolution.

Here we give examples of predictability and foreseeable futures for three areas of science: physics, biology, and economics (Faber and Proops 1993). We define the genotype for physical systems to be the 'laws' and fundamental constants (e.g. the gravitational constant, Planck's constant). These seem to be unchanging over time; that is, physical systems exhibit no genotypic evolution. We therefore define the foreseeable future of physics to be indefinite. In biology we observe genotypic evolution as the emergence of new species, something which occurs relatively infrequently. Thus for biological systems the foreseeable or predictable future will probably be of the order of millennia. In the economic systems of modern societies the emergence of novelty through technical innovation occurs very frequently, limiting the foreseeable or predictable future to a few years or decades. Below we will explore the implications of these different foreseeable futures, of physical, biological and economic phenomena, for decision-making about society–environment interaction.

ECONOMIC EVOLUTION AND PREDICTABILITY: THE ECONOMY AND THE ENVIRONMENT: *EX ANTE* AND *EX POST*

In the light of the above discussion on the predictability of phenotypic and genotypic evolution, what can we say about economic evolution in the long run?

First, we need to distinguish between economic evolution that is about the realisation of potentialities (phenotypic evolution) and that which involves the alteration of potentialities (genotypic evolution). Now the former is relatively easy to deal with; indeed, the great bulk of economic analysis is of this sort. Even that which passes as 'evolutionary economics' is, more often than not, about phenotypic evolution. (See, for example, Nelson and Winter's (1982) models of technical progress, in Chapter 9 of their book. The techniques are all given in their model, and only need to be searched for.) The reason that this method of economic analysis is popular is simple: phenotypic change is amenable to theorising, modelling, and prediction. If one knows the potentialities of the system, then it is not too hard to construct models of:

1 long-run equilibrium outcomes (i.e. static general equilibrium analysis);
2 comparisons between static outcomes under slightly modified model parameters (i.e. comparative statics);
3 dynamic behaviour when away from equilibrium (i.e. disequilibrium models and dynamic analysis).

However, if the potentialities of the system alter, it is very difficult to see how one can theorise about, and model, the nature of such changes. One can, of course, theorise about the sources of such change (e.g. growth of scientific knowledge, 'learning by doing', R&D). But just what the nature of new techniques will be is, by its nature, not something one can model and predict.

We can summarise this by invoking the familiar economic concepts of *ex ante* and *ex post*. If we can try to understand something *ex ante*, we are, in effect, trying to look into the future, to predict. If we are considering something only after it has happened, then we are exploring that happening *ex post*.

Now phenotypic change can be examined *ex post*, and that information used for theorising and prediction; i.e. also for consideration *ex ante*. However, genotypic changes are amenable only to *ex post* analysis. We can hope to understand why new techniques have come into being, but we are unlikely to be able to say anything useful about what new techniques (in a radical sense) will come into being in the future.

We now turn to economy–environment interactions; in particular we shall assess the predictability of environmental changes. To this end a useful distinction is between the invention of a new technique, and the commissioning of that technique throughout the economy. This process of commissioning a new invention we term innovation. It involves not only the immediate effect of a new technique, in the sense of making a new or improved good available to consumers, but also the ramifications of the new technique throughout the economy, and inevitably on the environment too. For example, the invention of the internal combustion engine led to the development of air transport, the building and surfacing of roads, and the setting up of networks of fuel stations. Further, it increased the demand for a natural resource, and it has generated pollution, etc. Thus we can imagine a temporal sequence as follows:

Invention → Innovation and production → Natural world

However, the impact on the natural world will generate further economic constraints that may lead to the search for new inventions. For example, the spread of iron-making with charcoal in the late Middle Ages in Britain made wood scarcer and more expensive. This led for a search for a new technique of iron-making, using coal instead of wood. Eventually such a technique was developed, by Abraham Darby in 1709, who substituted coke, derived from coal, for charcoal. Thus the constraints of the natural world may generate new inventions, so closing the cycle above.

The interaction between the economy, the natural environment, and the introduction of new techniques, can be represented by a diagram which shows us those changes in the world that can be explored *ex post* only, and those which can be explored *ex ante* and *ex post*. This set of relationships is shown in Figure 10.1. The right-hand side of the diagram shows that the commissioning of new techniques of production has an impact on the natural world, through resource use and pollution. To a reasonable extent, once the technique and its effects are understood, this changing set of relationships is predictable. The occurrences on the right of the diagram can be understood both *ex post* and *ex ante*. The left-hand side of the diagram shows that environmental constraints may give the incentive for new techniques to be sought; that is, invention may occur. If this invention does occur, it will constitute novelty, which is unpredictable. Therefore, the occurrences on the left of the diagram can be understood only *ex post*.

This diagram tells us that the analysis of the relationships between the environment and the economy in the long run cannot be made according to one approach only. We must recognise that while standard economic analysis, or some other variants, are applicable to the right-hand side of the

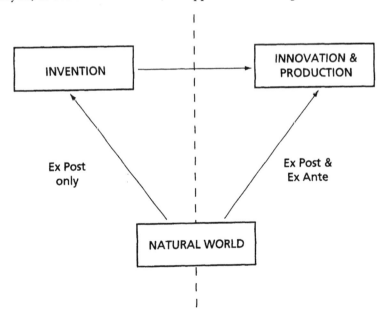

Figure 10.1 An iterative interaction between invention, economic activity and the natural world

diagram, the left-hand side cannot be explored that way. On the right-hand side we have predictable phenotypic changes, while on the left-hand side we have unpredictable genotypic changes. On the left-hand side one is going to have to take a more historical (i.e. *ex post*) approach, and any policy suggestions for that side of the diagram cannot be firmly based in theory. Rather, they are more likely to be of the type 'be careful, and try to keep your options open'.

ECONOMIC ACTIONS AND THE NATURAL WORLD: THE TOTAL IMPACTS LIFETIME

The four types of time period we have discussed so far all refer to decision-making. As well as these four notions, we now want to introduce the concept of a 'total impacts lifetime', which is not directly related to decision-making, but rather is a consequence of it, in interaction with the natural world.

By a 'total impacts lifetime', we mean the period of effect of an economic action upon the natural world. For example, imagine that it is decided to build a nuclear power plant. First, the power plant will be built and used. Once its useful life is over, the power plant has to be decommissioned, and the resulting nuclear waste has to be securely deposited. The whole period of time it takes from the commencement of building to the last damaging effect of the consequent nuclear radiation we define to be the corresponding total impacts lifetime of nuclear power generation. However, this description assumes there is no ignorance concerning the consequences of economic action on the natural world. Therefore, the assessment of the total impacts lifetime prior to the full effects becoming known we call the *ex ante* total impacts lifetime. In actuality, it is likely to be the case that there are other, unanticipated consequences, so that the total impacts lifetime *ex post* will generally exceed the total impacts lifetime *ex ante*. For example, the *ex ante* total impacts lifetime for refrigerators was considered small, as they are easily disposed of when no longer useful. We now know that a consequence of disposing of refrigerators is the release of CFCs, with the consequent long-run damaging effects on the ozone layer.

The relationship between the (*ex ante* or *ex post*) total impacts lifetime of an economic action and the above four time concepts is not very simple, for one cannot determine them in advance because of the existence of risk, uncertainty, and ignorance (cf. Faber *et al.* 1992). We are, therefore, able to make remarks only on the general nature of these relationships. First, one would expect that the total impacts lifetime for an economic action would exceed the lifetime of the corresponding capital. This is because the effects of the economic action on nature would generally be caused by the operation of the capital goods, and continue for some time after the capital goods have ceased to be used. By the same reasoning, the total impacts lifetime would also exceed the *telos* completion time.

However, in principle the total impacts lifetime could be longer or shorter than the entrepreneurial time horizon of the corresponding decision-makers. Of course, problems of environmental degradation are likely to result when the total impacts lifetime exceeds the entrepreneurial time horizon, as then foreseeable negative outcomes are ignored. As noted above, the *ex post* total impacts lifetime will generally exceed the *ex ante* total impacts lifetime, so that it is possible that even though the entrepreneurial time horizon is longer than the *ex ante* total impacts lifetime so that environmental effects can be taken into account, the *ex post* total impacts lifetime may still exceed the entrepreneurial time horizon, so unintended environmental effects occur.

Finally, the differentiation between the *ex ante* and *ex post* periods of impact may reflect the emergence of novel effects on the natural world, and the possibility of the emergence of novelty would be one factor in affecting the assessment of the foreseeable future for subsequent economic actions. That is, if it is recognised that economic actions often have longer *ex post* time spans than *ex ante* time spans, this may lead to a recognition of the emergence of novelty, and hence to recognition of the limits to predictability for economic actions in general, with consequent effects on the assessment of the foreseeable future.

Our example concerning the building and decommissioning of a nuclear power plant illustrates that even the *ex ante* total impacts lifetime is, strictly speaking, indefinite. However, the impacts decrease over time. For operational purposes it is, therefore, expedient to approximate the total impacts lifetime by a kind of radioactive half-lifetime concept. This concept has to be formulated in such a way that impacts decay to negligible proportions after that time. However, this approach can be employed only in those cases where the impacts decrease. If there exist positive feedbacks, as for the greenhouse effect, then the total impacts lifetime will increase and not attenuate.

UNDERSTANDING ECONOMIC ACTION AND ENVIRONMENTAL CONSEQUENCES

We now return to the main issue of this chapter, the relationships between economic actions and the environment, particularly with regard to time and predictability.

Relationships between the time concepts

We can now examine the nature of economic actions, using the terminology developed above, and relate this to environmental issues. We remind the reader that we have defined the following time concepts relating to the achievement of a certain goal or *telos*:

- *telos* completion time (related to the existence of a *telos*);

- capital lifetime (concerning the destructibility of capital);
- entrepreneurial time horizon (the period over which decisions are made);
- foreseeable future (the period over which no genotypic change occurs);
- total impacts lifetime (the period of the effect of an economic action on the natural world).

There exists an essential difference between the first four time concepts and the fifth concept, the total impacts lifetime. The former are consciously and actively taken into account in every decision-making process, because they all refer to achieving the corresponding chosen *telos*. It is possible to relate causally the economic effects of the first four types of period to the corresponding decision makers. Hence, all the revenues, costs, and profits can be ascribed individually to the decision-makers. This implies that a system of responsibility originating from the economic effects of the decision-making can be established. This ascription in turn is one of the pillars of a decentralised market economy in particular, and liberalism in general.

Turning now to our fifth concept, the total impacts lifetime, we see that it is not consciously and actively taken into account in the entrepreneurial decision-making process, but rather may be ignored, as discussed above. Thus there does not exist a comparable system of responsibility. It is therefore no surprise that the total impacts lifetimes of projects are not incorporated into standard economic analyses. From this it follows that, from the perspective of our time concepts, we see a dichotomy in responsibility for economic actions, concerning the purely economic effects and the impacts on the natural environment. This dichotomy is so fundamental that it may severely question the sustainability of the modern economic system of the Western world, and thus its basis, liberalism.

We also noted that the following relationships must hold between these various times concepts.

1 In general, the *telos* completion time is smaller than the capital lifetime of the equipment which is employed for production.
2 The entrepreneurial time horizon is at least as long as the *telos* completion time. If it were not, the entrepreneur would dismiss the project.
3 The entrepreneurial time horizon is generally longer than the capital lifetime. It is unlikely that decision-makers will often invest in longer-lasting (and more expensive) capital goods than is necessary to fulfil their objectives within the period of their concern (i.e. entrepreneurial time horizon).
4 The foreseeable future is longer than the *telos* completion time. This is because a *telos* cannot be reasonably formulated over a period greater than that over which predictions can be made.

5   The foreseeable future is usually at least as long as the capital lifetime, as it would be irrational to construct capital goods which lasted longer than the period over which predictions regarding their usefulness can be made.

6   The foreseeable future is at least as long as the entrepreneurial time horizon; no economic agent could sensibly make plans over a period greater than that which can be predicted. (This may be modified for large corporate players, who will back a number of alternative strategies in the face of uncertainty on the assumption that the bundle will include one or some which will pay off.)

7   The total impacts lifetime is at least as long as the *telos* completion time, as impacts will occur after the *telos* is achieved.

8   The total impacts lifetime is at least as long as the lifetime of capital, for the same reason as in 7.

9   The differentiation of the *ex ante* and *ex post* periods of impact may reflect the emergence of novelty, and thus influence the assessment of the foreseeable future.

The statements 1 to 8 are summarised in a systematic way in Table 10.1.

Table 10.1 The relationships between the five time concepts

	1 *Telos* completion time	2 Capital lifetime	3 Entre- preneurial time horizon	4 Fore- seeable future	5 Total impacts lifetime
1 *Telos* completion time		<	<	<	<
2 Capital lifetime			<	<	<
3 Entre- preneurial time horizon				<	
4 Foreseeable future					

Economic action and the environment

We first recognise that there are innate limits to the predictability of systems, be they social or natural. As discussed at length above, these arise from genotypic evolution, which generates novelty. Thus any rational choices made by agents will take place within the (generally acknowledged) limits of knowledge and predictability. Next, economic agents have entrepreneurial time horizons concerning their decision-making, which explicitly ignore eventualities that occur after this period. This is distinct from the limits to predictability mentioned above. Predictable outcomes may still be ignored.

What are the implications of economic actions for the natural environment? To assist in this analysis, we offer a stereotypical description of how the formulation of a *telos* may lead to environmental problems. First, a *telos* is established. Focusing on this *telos* will lead to the ignoring of other consequences of the pursuit and attainment of the *telos*. Some of these ignored consequences will be environmental in impact, such as the generation of pollution during the corresponding entrepreneurial time horizon. It should be stressed that this pollution need not be unpredictable.

Second, the decision-maker will seek to attain the *telos* within a certain entrepreneurial time horizon. Events subsequent to the time horizon will be ignored. Thus a second category of possible environmental effects may occur: those which are consequent to the attainment of the *telos*, but which lie beyond the time horizon. Issues of depletion of resources and sustainability are of this type.

Next, to attain the *telos* within the entrepreneurial time horizon, capital goods must be established. A consequence of the extended life of capital goods is that if, part-way through the process of achieving the *telos*, there is some unforeseen need to alter the *telos* in some way (e.g. because of environmental legislation), there is the necessity to alter the nature of the capital goods employed. Since capital goods take time to construct and bring into use, there will necessarily be a time delay between the *telos* being altered and the modification of the means to achieve the *telos* (Faber and Proops 1991). Thus if the achieving of a *telos* leads to environmental problems, as mentioned above, and if there are policy decisions to reduce their negative effects, there will be a considerable time lag between, say, legislation being enacted and the full adjustment of the production programme towards the new, modified, *telos*. For example, under environmental legislation one would expect to see a move away from coal-burning electricity generation towards gas-burning and wind- and wave-powered generating stations. This modification of the corresponding capital stocks may take several decades to complete.

Finally, although one would anticipate, as argued above, that the *telos* sought would be achievable within the foreseeable future of the process (i.e. before novelty can occur), it is still feasible that there will be novel outcomes

214

encountered in seeking the *telos*. For example, the *telos* of food storage has largely been achieved through the use of refrigerators using CFCs as refrigerants. These refrigerators were designed and constructed on the basis that novel and superior methods of refrigeration (or food storage) were not expected to be found (as they have not been). However, novelty of a different type has occurred: we have recently discovered that CFCs cause the depletion of the ozone layer, with its damaging environmental effects. That is, the *ex post* total impacts lifetime of refrigerators considerably exceeds the *ex ante* total impacts lifetime.

Thus one can establish a series of environmental problems associated with the temporal nature of a teleological approach to human behaviour. First, attention on the pursuit of the *telos* may ignore most side effects, even though these effects are apparent to all. Such ignored effects may well be environmental in nature (e.g. air and water pollution resulting from metal manufacture). Second, the existence of an entrepreneurial time horizon for decision-makers will mean that many future outcomes of present actions will be ignored. The ongoing processes of soil erosion, forest depletion, and overfishing of the oceans are examples of such negative environmental effects. Third, the nature of capital goods, being specific in their use and requiring time to construct, means that there is considerable 'inertia' in modern production systems. A consequence is that even when environmental legislation is introduced, it may take many years before it can be fully effective. Fourth, novel types of environmental problems can emerge, associated both with new means of production and with old ones.

CONCLUSIONS: DECISION-MAKING CONCERNING THE ENVIRONMENT

If we accept that human behaviour is teleological, and this teleology is circumscribed by the existence of decision-makers' entrepreneurial time horizons, the requirement of capital goods to achieve *tele*, and the possibility of the emergence of novelty, what should be the form of decision-making concerning the environment?

First, the narrowness of vision that seems to come from teleological behaviour needs to be recognised. A consequence of this is that even with completely assigned property rights and the existence of appropriate markets, there may still be considerable 'intertemporal external' environmental effects. These will require some form of legislative action to control them. Second, the existence of entrepreneurial time horizons for decision-makers demands that long-term environmental damage be dealt with by governments, acting with appropriately longer time horizons than individuals or firms. Problems of sustainability and resource depletion demand long-run legislative controls. Third, the introduction of environmental legislation will always be followed by a lag before the effects of the legislation become apparent, because of the need for capital restructuring. Fourth, the possibility of the emergence of new and unpredictable environmental problems will always remain. On the other hand, the continuing emergence of new inventions may (or may not) offer solutions to present environmental problems, though they may also generate new ones in the future. This is much the most difficult problem of decision-making we need to confront. An explicit recognition of the unpredictability of many phenomena tells us that there can be no way of defining now 'correct decisions' concerning future outcomes. Hence the likely emergence of new and unpredictable environmental problems resulting from economic activity can only tell us to be cautious in introducing new technologies. In particular, it tells us to seek flexibility of response to unforeseen problems; we should try to avoid locking ourselves into economic processes that may prove more harmful than beneficial. (However, it is important to recognise that it is easy to get locked into what subsequently are seen to be inferior technologies because of the complex system of support that grows around them, and because of increasing rather than diminishing returns.)

On the other hand, the possibility of new inventions allowing us to circumvent current environmental problems suggests that we pursue the search for new technologies vigorously. Of course, this to some extent contradicts the need for caution in introducing new technologies mentioned above. Perhaps the overall message regarding the emergence of novelty and the environment is that we should plan for flexible responses, while simultaneously seeking new methods of production, which themselves embody flexibility for the future. The biggest problem in this is knowing to what extent private entrepreneurs can be expected to behave in this manner, or the extent to which public authority has to be reasserted in this age of liberalism.

NOTES

1   A 'causal mechanical' interpretation of Hamilton's principle is also possible. Teleological and causal descriptions of processes can be seen as complements, rather than exclusive alternatives (Bertalanffy 1950).
2   Faber et al. (1995a) argue that the interactions of species in an ecosystem can also be described as goal oriented (i.e. teleological).
3   The telos completion time corresponds in economics to the 'absolute period of production' (cf. Faber 1979, p. 21). We recognise that the *telos* completion time may (and often does) change historically, according to the technology and resources available.
4   An excellent overview of the problems of discounting is given by Lind (1982).
5   The problem of how planning can extend over an indefinite time when there is a finite time horizon is easily dealt with by the use of 'rolling plans', where actions are determined by a series of finite but overlapping decision periods. For a discussion of the theory of such 'rolling myopic plans', see Faber and Proops (1993, Chapter 10); for a survey on the wider nature of sequential planning, see

Schmutzler (1991). For a discussion of how the length of the time horizon affects the implementation of economic plans and, in particular, the innovation of new techniques see Faber and Proops (1991).

6 Arguments for the necessarily finite lifetime of capital goods can be based on the entropy law of physical science (Georgescu-Roegen 1971; Faber *et al.* 1995b).

7 For a much more detailed presentation see Faber and Proops (1993, Chapters 2 and 3).

REFERENCES

Arthur, W. B. 1989. Competing technologies, increasing returns and lock-in by historical events. *Economic Journal* 99: 116–31.

Bertalanffy, L. von 1950. The theory of open systems in physics and biology. *Science* 111: 23–7.

Faber, M. 1979. *Introduction to Modern Austrian Capital Theory*, Heidelberg: Springer Verlag.

Faber, M., and Proops, J. L. R. 1991. The innovation of techniques and the time horizon: a neo-Austrian approach. *Structural Change and Economic Dynamics* 2: 143–58.

Faber, M., and Proops, J. L. R. 1993. *Evolution, Time, Production and the Environment*, 2nd edn. Heidelberg: Springer-Verlag.

Faber, M., Manstetten, R., and Proops, J. L. R. 1992. Humankind and the world: an anatomy of surprise and ignorance. *Environmental Values* 1: 217–41.

Faber, M., Manstetten, R., and Proops, J. L. R. 1994. *Knowledge, Will and the Environment*. Discussion Paper no. 205, Department of Economics, University of Heidelberg.

Faber, M., Manstetten, R., and Proops, J. L. R. 1995a. On the conceptual foundations of ecological economics: a teleological approach. *Ecological Economics* 12: 41–54.

Faber, M., Niemes, H., and Stephan, G. 1995b. *Entropy, Environment and Resources: An Essay in Physico-economics*, 2nd edn. Heidelberg: Springer-Verlag.

Georgescu-Roegen, N. 1971. *The Entropy Law and the Economic Process*. Cambridge, MA: Harvard University Press.

Hicks, J. R. 1973. *Capital and Time: A Neo-Austrian Theory*. Oxford: Clarendon Press.

Holling, C. S. 1973. Resilience and stability of terrestrial ecosystems. *Annual Review of Ecological Systems* 4: 1–24.

Lind, R. C. (ed.). 1982. *Discounting for Time and Risk in Energy Policy*. Baltimore: Johns Hopkins University Press.

Mirowski, P. 1984. Physics and the marginalist revolution. *Cambridge Journal of Economics* 4: 361–79.

Mishan, E. J. 1969. *The Costs of Economic Growth*. Harmondsworth: Penguin.

Nelson, R. R., and Winter, S. G. 1982. *An Evolutionary Theory of Economic Change*. Cambridge, MA: Harvard University Press.

Schmutzler, A. 1991. *Flexibility and Adjustment to Information in Sequential Decision Problems*. Heidelberg: Springer-Verlag.

Solow, R. M. 1985. Economic history and economics. *American Economic Review* 75: 328–31.

Tietenberg, T. 1994. *Environmental Economics and Policy*. New York: HarperCollins.

Waldrop, M. 1992. *Complexity: The Emerging Science at the Edge of Order and Chaos*, Harmondsworth: Penguin.

11

INDIA, DEVELOPMENT AND ENVIRONMENTAL CHANGE

Graham Chapman

INTRODUCTION

Many of the chapters of this book have looked at the rate of past environmental change, and at potential rates of change – or more likely at potential rates of changes in the rates of change – of the environment in the short- and medium-term future. Many have also looked at how some rates of change – for example plant migration – can or cannot keep up with, for example, changes in climate. Some have also looked at the rates of change in societies which have been confronted by problems of climatic change. In Chapter 3 there is considerable discussion of the failure of the Vikings in Greenland to adapt to what for them appeared to be worsening climatic conditions, although other cultures survived in even 'worse' circumstances. What such studies suggest is that there is no way in which we can talk meaningfully of a rate of environmental change without talking about its impact on something else. From the perspective of human society it is the direct and indirect impacts on society that matter most in determining what is a critical rate of change – a point which Max Wallis also heavily underscores.

Obviously human society does not change only as a result of external stimuli represented by environmental change: most historical studies of society and economy stress the changes which have occurred as civilisation has 'progressed' – changes in the concepts of the state, the concepts and methods of government, the acquisition, organisation and transmission of knowledge, the patterns of settlement, the relations of production, and the advance of technology. At most stages such studies have included discussion of some sort about relationships with the environment, be it in terms of types of agriculture, the use of water, or the extraction of energy. However, it is only recently that there has been widespread concern over environmental impacts at a large scale – i.e. those threats such as the greenhouse scenario – which represent unwanted feedback from a changing environment with the potential to affect much of mankind. It is natural that the debate over these

218

issues should concentrate on the scale and timing of such threats, but in doing so too often the presumption is made that human society will have to adopt radical changes quickly to avert the perceived catastrophes. The presumption becomes a kind of utopian imperative, because the threats from the doomsday prophets suggest that there are no alternative courses of action. But such an approach does not address the questions which have interested many previous generations of scholars, and many current development specialists, namely what it is in human society itself that controls the rates of social and economic change independently of environmental considerations.

In a rather simple way we can suggest that there are two ways we can look at society–environment interaction. The first is by starting with observations and theories about the behaviour of the environment, and then adding to this base our understanding of society. The second is to consider our knowledge of, and theories about, the development of human society, and then to add to that our understanding of its interaction with the environment. Naively we might expect either way to come to the same results, but in practice this does not happen, and the results are actually very divergent. The reasons for this are not too hard to discern, and are tied up with the different methods of argument and explanation which are used by the sciences and the social science/humanities. These are summarised in Table 11.1. If we start with theories of the environment based in the left-hand column, we know that the principles of physics, which do not vary in time or space, can be used to construct a general circulation model (GCM), and the proponents of the model will argue for its validity because they believe that it is based on demonstrable and unchanging physical laws. If we add the biological realm things get a little more difficult, because evolution in part has a historical explanation, and so is not necessarily explained only in reductionist terms. Species within ecosystems co-evolve. The best attempt to put these two approaches together is probably exemplified by James Lovelock's inspired but still controversial Gaia hypothesis: that biological life has co-evolved with the atmosphere. How does one go one step further, and add the human world? The answer is that it is possible to do so only if we limit the concept of 'people' to the reduced forms shown in the bottom line of the left-hand three columns – in short, people as physical and biological artefacts. It is not possible to model the properties of the last column in any predictive sense. In any case, the world of thought is reflexive – that is to say, we are aware of our own thought processes. Reflection of this sort means that this world is self-referencing, with all the logical traps that this immediately implies. If, for example, a GCM were augmented to include this human realm, then it ought to be able to model the fact that if its predicted outcomes are 'bad', then human activity could avert such outcomes, in which case its predictions would be wrong, and therefore changed so that they were not bad, in which case . . .

It is therefore fairly easy to see that if we start with environmental models society will be tacked on in very simple ways. Mostly, as we have just noted above, society will not actually be modelled along with the environment; rather it will be exhorted to heed the oracle and act accordingly. Society does not exist analytically, but only as an undifferentiated 'something' (usually 'mankind') which has to implement the prescribed normative action. The complexity of human society is wished away, and no account is taken of the variety of cultures in a variety of local environments, either of their value systems or of their stage of 'development', which will affect profoundly the

Table 11.1 Modes of understanding and explanation in the physical, biological and social worlds

	Physical/ chemical world	*World of human artefacts*	*Biological world*	*World of conscious human thought*
	Complication	Complification	Complexity	Complexification
Under-standing	Nil or cosmological	Social and historical context	Environmental	Consciousness (philosophy, ideology, values)
Explanation	Reductionist Universal (statistical)	Reductionist Deterministic Certainty Universal Ahistorical Bounded	Holistic (and reductionist in part) Emergent Historical Evolutionary	Holistic Emergent Reflexive Uniquist Historical Unbounded Uncertain
Examples Irrigation systems	Kilometres of channels, of outlets, of regulators Sedimentation Salinity	Integrated design system	Ecosystem Plant growth Genetic diversity	Political mandate Crop localisation Protective/ productive Communication Perception Corruption
People	Numbers Required food supply Demographics	Robots	Patient of allopathic medicine	Personality Ambition Mood Aggression Altruism

Source: Chapman 1993.

way in which they can respond to 'external' challenges. So somehow, because the scientists of the North have produced good GCMs, everyone around the world 'ought to' participate in a universal system of mitigation.

If we start from the other end, by modelling society, then clearly we have to see how this has been done by those working in the social sciences and the humanities. The focal point is always on how people interact with people, to form the larger groups of cooperators and competitors which comprise society. For an early social scientist, such as Karl Marx, the essential model is derived from the atomic models successful in physics – namely that the people are like atoms with certain valencies, and can and do combine in certain ways, to produce feudal, capitalist or socialist societies, in which what varies is not the environment or the essential nature of mankind, but the structure of the relations of production and distribution. Most sociologists now deny the supremacy of the economic relations of production, and allow human beings to be more human. They look in finer detail at other patterns of cooperation and competition, in nuclear and extended family networks, in group behaviour in the workplace, and in patterns and styles of living among many other aspects of human society. The social anthropologists extend such studies to cross-cultural comparison. Thus although early social science attempted to find universal laws of history – and for Marx the 'progress' of society towards communism was inevitable – the patterns of explanation and understanding of most contemporary social science belong firmly in the last column of Table 11.1, where human aspiration and agency is given its full due. Alone among the social sciences, economics, as opposed to economic history, has tried to remain aloof from this complexification, and has persisted in a physics-based model (an approach criticised by Faber and Proops in Chapter 10) of an animal unknown on earth: well-informed Rational Economic Man. Post-modern approaches in the social sciences and humanities suggest that there neither is nor can be any single analysis or model of any social system – but rather a multiplicity of understandings and viewpoints, none privileged a priori over the other.

If the starting point for analysis and understanding is the interaction between people, how does one add the environment to the basic picture? If society and environment interact, then it ought to be the case that one does not simply add the environment, but shows how interaction with it affects the very structure of the relations within society. This is possible to some extent with 'primitive' societies studied by anthropologists, but since the industrial revolution and its attendant huge specialisation of labour, from the 'source' perspective few people appropriate anything directly from the environment themselves except for the air they breathe or consume with their vehicles. It is easy to see that the essential features of the structure and behaviour of society that we tend to study are actually far divorced from the environment. If we extend this logic far enough and hypothesise, as many do, a single world economy in which the specialisation of labour anywhere is seen only in terms of the one global economy, then this one economy plugs into one and only one global environment through a selected number of agents. These agents are the people who are listed as primary producers: the farmers, fishermen, miners, etc., who are very few in number in developed societies. Thus the complex model of society interfaces with a simplified, if not trivialised, concept of environment. One can see why as a reaction to this, and despite the fact that it is not actually necessary for the members of an environmentally conscientious advanced society to be individually responsible for all their environmental demands by direct interaction with it, some 'green' enthusiasts are driven to 'back-to-nature' styles of living, and other propagandists associate good environmentalism with 'local community' and sustainability with 'local carrying capacity'.

One discipline above all has made it its business to attempt to fuse theories which embrace the variety of society with variety in the environment, namely geography. In the first half of this century this attempted fusion produced the idea of regional geography – in which, despite the many beautiful and valuable descriptive accounts of the regions of the earth which were completed, both the social and the environmental aspects of different regions were simplified mostly beyond theoretical redemption. What this current chapter attempts to do is to approach environmental questions from an understanding of the specificities of a particular culture – India – to show how the rhythms and time-scales of change inherent in this society impact on its relationship with its own local (but subcontinental) environment, and also to show why global concerns are of less importance. To make some of the statements clearer, from time to time certain aspects are contrasted with British equivalents.

INDEPENDENT INDIA, NATIONALISM, AND THE IDEA OF DEVELOPMENT

Whenever we look out at the world around us we do so on the basis of unexamined assumptions. By this we mean not that these assumptions are never examined, but that they are not examined in the shorter-term acts of experiencing our surroundings. The unexamined assumptions come at different hierarchical scales. We hold assumptions about trivial things, such as that we will have breakfast tomorrow morning, that the post will arrive. We assume less trivial things: that the students in class this morning will not have a mass shoot-out in which they kill each other. The whole of communication depends on shared assumptions. For example, news presenters in the UK assume that most of the audience will have some idea of what NATO means, and that they can distinguish between local and parliamentary elections. It is assumed by TV presenters in Britain that the audience know that Britain has not been invaded for nine hundred years, that it has a long history of democratic government, that it is a post-imperial middle-ranking power, which is quite wealthy and technically advanced in global terms, but which in relative terms has slipped down international rankings quite a lot since the Second World War, and that it has an ambivalent attitude towards the rest of Europe – 'the Continent'. There is an assumption that, for all its problems, Britain is not a 'developing' country.

It is impossible to stress enough how different are the lenses through which India views itself and the world. India is not even confident yet that India really exists. For the same nine hundred years during which Britain has not been invaded, much of India has been under imperial rule by dynasties originating from outside India. This includes the 300 years of the Afghan Sultanate of Delhi from AD 1100 onwards, the 300 years of the Moghuls from 1500 to 1800, and the 150–200 years of the British Raj, ending in Independence in 1947. India was not seen even by these rulers as a country – it was seen as an empire of nations in itself. The British were the first to bring the whole of South Asia under one hegemony – but they did not unite it as one country, since it was a hotchpotch of provinces treading slowly towards home rule, and a collection of monarchical authoritarian states (the Princely States of the Maharajahs and Nawabs) under British protection. At Independence in 1947 the ramshackle edifice fell apart (the finger of blame is pointed in all directions) and gave birth to West and East Pakistan (now Bangladesh) and what became the Republic of India. Well over a million people died in communal massacres, and nearly 20 million fled their homelands – a sort of ethnic cleansing, spontaneous but also instigated, such as seen more recently in former Yugoslavia. Therefore since 1947 the central government of India has had as an absolute priority the instillation into the minds of each and every one of the citizenry of the republic that they are first and foremost Indian. It has had remarkable success – something from which Europeans have much to learn. But it is not total success, and the dark forces of Hindu–Muslim communalism, linguistic communalism, and regional secession remain. The running sore of Kashmir is evidence of this. But every day the schoolchildren all over India sing the national anthem a hymn to all the states and regions of India. This may be called indoctrination, or it may be called socialisation.

The first step of the government of independent India was therefore to cultivate the idea of secular Indianness – but that alone is not enough to legitimise the union (i.e. federal) government. This central government must do something for its people; in modern political parlance it must pursue a 'great idea', so that they can all see how they as Indians are benefiting from *swaraj* – self-rule – in independent democratic India. It is also a truism that there would be greater cohesion among the citizenry if they felt themselves to be some kind of 'us' against some kind of 'them'. 'They' were clearly the industrialised, imperialist countries, one of which had ruled India for the last century and a half, and in the view of the Prime Minister (Jawaharlal Nehru) left much of India sunk in backwardness, ignorance, poverty, and superstition. Nehru believed that for the country to progress, planning by the government was essential, and that 'planning was science in action' (Vasudeva and Chakravarty 1989, p. 417).

In short, the government would seek to 'develop' India, and readily acknowledged its place among the 'developing nations'. It took inspiration from the rapid development (and in the early years it was rapid) of the Soviet Union, and copied the idea of five-year planning and public sector dominance in the commanding heights of the economy. India turned its back on 'economic dependency' and became a closed economy. Public sector investment was poured into what turned out to be inefficient heavy industry – but the rural sector was not forgotten. Here the effort was as much administrative as financial. Partly this involved land reform and land redistribution to varying extent in different states, partly it involved major innovations in the structure of rural government. The structure of government that the British had inherited from the Moghuls, modified and then passed on to independent India, reached down to district level – a district being a subdivision of a province and commonly having one or two million inhabitants. For taxation and police purposes there were subdivisions of the district – but the district and its District Officer were paramount. In sweeping reforms in the late 1950s, villages were grouped into new blocks, about 100 villages at a time, to implement a programme known as Community Development. Each block was headed by a Block Development Officer, who oversaw Village-Level Workers. Nor did technology pass rural areas by. India accelerated the development of large-scale irrigation, and began the construction of huge new dams – the temples of modern India as Nehru called them. The point is that the word 'development' intrudes everywhere that society and government intermesh. And the idea of science intrudes everywhere that development and society intermesh: new seeds, new fertilisers, new health clinics, and massive electrification programmes.

In the 1980s the small state socialist economies of Africa and some of the large economies of Latin America all succumbed to demands from the World Bank and the IMF for liberalisation and restructuring. In India's case, her size and the fact that comparatively speaking she has not had such a large overseas debt has meant that external pressures have not worked either so fast or so simply. But the realisation by Indian leaders and by different sectors of the new entrepreneurial classes that growth has been slow, that to be modern India's industry has to compete with that of the rest of the world, together with pressure from GATT (now the World Trade Organisation) and its Northern bosses, has indeed led to a policy of liberalisation and increasing openness. Now international investment is more likely courted, not spurned. The policy is not, however, completely firmly set, since vested interests in India are hurt by it. A reversal of policy seems unlikely, but this does not mean that the need for the policy is not widely resented. The majority of India's intelligentsia have never been outside India – for obvious economic reasons. They have little exposure to the 'reality' of the North, and see it only through what is on TV and through the press. They have a strong feeling that 'they' (the Northern industrialised nations) are dictating to 'us' (Indians, but representatives of the poor South). The North tries to dictate – over nuclear non-proliferation, over the patenting of genes ('our' neem, a tree which produces a natural pesticide, is being exploited by the 'multinationals'), over trade liberalisation, over the terms and costs of the transfer of technology. 'If they want our industries to be clean, they should give us that technology' was a remark made often in a seminar I attended in Shimla in May 1995. All this is of course refers to a perceived 'unfairness' and reveals a strong suspicion of neocolonialism at work. In environmental terms the perceived unfairness is also allied to the widely held view that actually it is the North that is the environmental disaster, each individual consuming thirty times the resource needs of an average Indian, with the inevitable result that pollution by the North is horrific. (Associated with this is the perception that in the North social life has been debased and all family values abandoned.) The article published in *The Tribune* from Chandigarh, shown here as Figure 11.1, is a recitation of some of these themes. My point is not the truth or otherwise of these views. It is that again we come to the unexamined assumptions. One of them in India is that a major part of India's predicament is that the global system disempowers the country, and works against it.

Of course there have always been dissenting critics in India of the whole modernist project, some of them decidedly Gandhian in outlook. Mostly they have objected to big cities and industrialisation as an avoidable evil, something that is seen to destroy family life, and that has Kali Yuga quality (a Hindu concept that we are now in an age of destruction – see Chapter 1) about it, like Blake's 'dark, satanic mills' in England. They stress rural development with appropriate technology, and institutional change to remove inequality in society. In this view, technology follows rather than leads, and institutions are built bottom-up. The Gandhian approach is revealed, in modern terminology, to be very environmentally aware, although during Gandhi's lifetime the environmentalist movement had not independently reached a peak which he could capture for his own purposes. The Gandhian dream is always honoured in India, but honoured more in the breach than the observance. The idea is kept alive as a kind of national conscience: utopian communities supported by private donations operate in all parts of India, but they show no sign of becoming the dominant paradigm.

THE INDIAN ENVIRONMENT

By common usage, India is referred to as a subcontinent (which technically includes of course Pakistan, Nepal, Bhutan, and Bangladesh). Modern plate tectonic theory explains why it is a subcontinent. Most of the southern landmasses (that is, south of the line which runs from the Persian Gulf

The Tribune, Chandigarh, 21 May 1995

Sunday Reading
Running to Stay in Place

By Gagan Dhir

Why is human hunger widespread in India but not in China, where there is half as much arable land? The pressure from a growing population is an obvious symptom but not the main cause of poverty, hunger and environmental degradation. The biggest contributors to the deteriorating environment are nations with stable or decreasing population.

... One quarter of the world's people living in developed countries consumes 75 per cent of the world's energy, 79 per cent of all commercial fuel and 85 per cent of all the wood. Each person in Europe, North America, Australia and Japan consumes, directly and indirectly, enough grain to feed four Indians. The average Swiss consumes 40 times more resources than the average Tanzanian.

Because they consume more, developed countries pollute more. Paul Shaw of the UN Population Fund calculated that in 1993 rich countries were producing approximately 1.9 tonnes of waste per capita. Poor countries were generating just 0.2 tonne per person and as little as 0.08 tonne in rural areas.

Says Shaw: 'This means that the biggest contributions to environmental degradation – measured by waste generation – comes from countries where population growth has been stable, if not declining.'

Note: the article never gets around to answering the question it poses at the beginning about hunger in India.

Figure 11.1 Article from *The Tribune*, Chandigarh

through the Mediterranean to the Caribbean) formed one large land mass known as Gondwanaland. About 200 MA (million years ago) Gondwanaland broke up, parts drifting off to form Antarctica, a part to form Australia, the major part remaining as Africa; but another chunk, what is now the Deccan block of peninsular India, drifted north across the Arabian seas. About 80 MA it first bumped into the southern flank of Asia. The impact of this collision is clearly seen on a world map: a rim of massive wrinkled mountains to the west, north (Himalayas) and east of India. The process continues: the Deccan block is still pushing into Asia, at the rate of 6 cm a year, and the mountains are still going up at the rate of 6 cm a year. This is the arena of the strongest of inter-plate collisions on earth.

India is most easily divided into three macro-geological zones: the young mountains of the north, the highest and most extensive on earth; the old Deccan block, this errant fragment of Africa; and the massive river flood plains, the biggest and deepest alluvial plains on earth. These components are locked into a causal chain together with the climate, which is extreme. The most significant feature of the climate is the extreme seasonality of the rainfall: for much of India this is restricted to the period of say three or four months of the monsoon. The monsoon is part of the annual movement of the inter-tropical zone of convergence, the same phenomenon that gives the Amazon and the Congo their rainfall, but in those areas the northward and southward movement of the rainy season is less dramatic. The massive Himalayas are high enough to inhibit the northward movement of the high-altitude subtropical jet stream in summer, until a sudden switch occurs sometime in late June. The monsoon winds that then flood over India are 6,000 metres deep, compared with the 2,000 metres of the East Asian monsoon, and the relative humidity is higher. When they collide with the Meghalaya block north of Bangladesh, they give Cherrapunji the highest rainfall in the world. And for much of the mountain front of India, rainfall is immense. Given this, and the height of the still lifting mountains, and given the often comparatively soft geological strata, the natural rate of erosion is the highest on earth. The rivers debouch onto the plains, laying the massive alluvial deposits, and finally in Bangladesh, where the Ganges and Brahmaputra are confluent, the rivers form the world's largest and most active delta. At peak flood the Brahmaputra in Bangladesh may be 40 km wide – wider than the Straits of Dover.

The wettest parts of India are in general the western coastal areas of the Deccan, the lower Ganges valley, and the Brahmaputra Valley (Assam and surrounding mountainous states). Only in the extreme southwest, in Kerala, is there a short enough dry season for the area to be called 'tropical moist'. In the other 'wet' parts the dry season extends between seven and nine months, and in the summer months of April to June, as temperatures climb to 40 °C and above, the country is desiccated. In the central Deccan and the upper parts of the Ganges, rainfall totals are less, and inter-annual variability greater. Most of these areas are drought-prone, and without irrigation agriculture is an extremely risky business. Finally, in Rajasthan and parts of Gujarat on the Pakistan border there is a real desert, the Thar desert.

The variation in river discharge is therefore also extreme. In the Ganges basin many of the south-bank rivers, massive rivers in the wet season, dry up completely in the summer. The north-bank rivers show almost equally great extremes of discharge, but they do not dry up in the hot season because of glacier and snow melt in the mountains. Thus neither the Ganges nor the Yamuna should cease to flow.

For perhaps nine months of the year, away from the coasts, there is very little wind. The weather is stable and predictable in a way which would warm any cricket lover's heart, and outdoor badminton courts are a frequent sight. In north India, Delhi for example, in the winter months of December and January local temperature inversions of the sort that give London a smog for a few days may persist for the whole two months.

Bombay and Madras, two port cities founded by the British, are coastal. Calcutta looks coastal, but is in fact as far inland as Birmingham, up a distributary of the Ganges delta which has been losing its river flow through natural causes over the past few centuries. All the other big cities are far, far inland. Some, like Delhi, are near a perennial watercourse, in this case the Yamuna; others, like fast-growing Bangalore, are in the drought-prone central Deccan.

The big cities mix industry, power stations, and housing in close proximity. They are overrun by two-stroke scooters and badly adjusted diesel trucks and buses. The use of soft coal, wood, and rubbish for cooking is common. For much of the year the particulate matter simply lands again in the same area from which it came. During the winter inversions, not even the gases escape outwards. These cities have some of the world's worst air pollution – a level unimaginable till experienced. They are short of water – particularly of course in the hot summer. Fewer than half the households may be connected to a public water supply, and certainly fewer than half have flush sanitation. Where would the water come from to flush the other half, even if the equipment were installed? The newspapers daily run stories on the water wars between the states of Punjab and Haryana and the union territory of Delhi – yet the urban areas continue to grow and grow. The abstraction rate of water from the Yamuna at Delhi has more than once reached the point where water is being drawn back upstream from what should be downstream sections – into which untreated effluent is poured. Urban rubbish collection systems are inadequate, and in many areas rubbish creates an eyesore, is a health and sometimes a fire hazard, and is an olfactory insult. But, as Indians are fond of pointing out, they produce far less per capita than westerners, proportionately much more of it is recycled (more of this later), and less of it is non-biodegradable. It is just that to the Westerner the small amount of plastic represented by billions of small, thin polythene bags seems to have managed to spread itself over every part of the country, town, and countryside, to the apparent unconcern of the populace. It is not just a visual offence: the plastic litter gets stuck in the inadequate drains and blocks them. At least in respect of air pollution it might be thought that rural areas are cleaner than urban areas, and for men they are. For women it is less true, since rural kitchens rarely have chimneys, and cooking fuel – often dung – is very smoky. Even now there are many villages with only meagre and contaminated supplies of surface water available in the summer months, increasingly liable to industrial pollution as small-scale industry spreads, although, with increasing electrification and tubewells, many now have much more reliable and healthier supplies than before.

As befits a subcontinent with such a diversity of environments, and such an idiosyncratic history – for 60 million years an island in the Arabian Sea – the natural biodiversity is very high. Naturally, much of India would be covered with forests, mostly tropical dry deciduous forest – meaning that the trees lose their leaves in the hot dry season, not the cold winter months as in Europe. But India's colonial history and recent urbanisation and population explosion have seen to it that many of these resources have been reduced to rump status. Perhaps a little more than 10 per cent of India remains forested, and only some of that forest is in any kind of pristine condition. Of course it is hard to get accurate data on forest cover, because the topic is beset with definitional problems about density, age, and variety of tree cover, and because the best of what is left remains intact mostly because it is inaccessible. But modern techniques of remote sensing are helping, and the government takes the issue seriously, without external prompting by the world community. In a country where perhaps half the population still rely on renewable energy for cooking – cow dung and wood being the commonest such fuels – forestry is not a matter which is left to the Forestry Commission, with the simple caveat that the public should enjoy rights of access for Sunday walks with the dog. Society and trees interact at every level. In the Bengal delta there is virtually no remaining natural tree cover, yet each hamlet is surrounded by, buried within, and shaded by trees and woody species such as bamboo. Each may have several uses: for boat-building, for furniture, for resins to proof fish nets, for fruit, for fodder for animals, for agricultural implements, and for lopping for fuel. This is a productive and in some sense biodiverse arboreal landscape. But it is not the natural bio-diversity, and it is certainly less diverse than the original forest.

The other great fear that inspires modern environmentalist apprehensions of rural India's fate is the fear of massive soil erosion and soil impoverish-ment through exhaustion and/or incorrect use of chemicals. We will not enter here the debate about the high mountains (see Chapman and Thompson (1995) for a fuller account of that), since natural rates of erosion there are high anyway, and actually the proportion of India's population in these areas is comparatively small. We are therefore concerned here more with the gentler slopes of rain-fed agriculture in the Deccan and the Gangetic plains, and with the terraced wet paddy lands (which may be irrigated or rain-fed depending on local conditions) of the same two regions. The stories about soil impoverishment through excessive fertiliser use mostly emanate from the most intensive farming area of India: the Punjab. The problem is that the truth of the matter is not known, the evidence anecdotal and repeated and generalised until it is a self-evident truth. As far as we are aware there has been no state-wide survey which carefully links the empirical with the theoretical. Since tropical and subtropical soils have a lower organic content than temperate soils and recycle nutrients faster, and have higher rates of bacterial activity, it is quite possible that some kinds of damage occur faster than with equivalent doses on English clay soils, say. But this remains conjectural at the moment. Soil erosion is often exemplified by the ravines and gullies of the badlands of the river Chambal and its basin. This, however, happens to be a late Pleistocene loess dump – wind-blown fine sedimentary material. The largest example of such a loess dump is in China, often described as the world's largest waste tip, an area of massive erosion and catastrophic slumping. These deposits are way out of equilibrium with current conditions, and it is their natural state to erode and gully massively. This is nature working long term, and man has little influence over the rates observed. Erosion on degraded forest land is quite clearly and quite often a problem, which in places is checked by reforestation. But massive erosion on good agricultural land is not common; it is too precious, and farmers are too careful. But then of course one can look at the formidable extent of brick pits surrounding urban India. These use local earth, and are never dug deep because little mechanisation is available, and because of monsoon flooding. Typically only the top 2 to 4 metres of earth is dug away. They are therefore very extensive – a kind of dug environment mirroring the built environment. But even here on close inspection the situation is not as obvious as it first seems. Given the depth of alluvial soil in many places, the mineral content of the revealed surface usually returned to agriculture is quite often better than the well-used soil that was removed. It does, however, have a lower organic content. Further, it is not unusual for the pit user to want to enhance the value of the land by digging the pits so as to enhance the possibility of irrigation. The results are as likely to beneficial as they are to be deleterious.

Indian agriculture has increased its yields through a combination of factors – new techniques, inputs, increasing the number of crops per year – but all of these mostly work only in combination with irrigation. There are many sources of irrigation water, but two dominate. The first is the large surface canal systems which lace many of the river plains, fed from river barrages, with the rivers themselves often fed from dams in the hills. The second major source is tubewells, which have exploded in number all over India in the last two decades. In much of the hard-rock Deccan, these are the only possible sources of water, but the Deccan rocks have variable and usually small aquifers. In many areas the pumping is lowering the water-table. In the Ganges plains the situation is far more complex. Until a few years ago most of the alarm – and even now a great deal of alarm – was expressed because the leaky canal systems were causing a rise in water-tables, leading to constant evaporation from damp soils, which turns them saline. On the other hand one reads in the press – but there are few academic studies of the causes – that tubewells even here are reducing water-tables to the point where irrigation is no longer possible. Farmers like tubewells because they get water when they want it, not when or if the canal system gives it to them. In the technical literature much is now written about 'conjunctive use' of surface and groundwater – but in practice there is little of this since the two systems are under different control. The groundwater reserves of the Ganges plains ought to be adequate for much more extensive exploitation than has so far occurred – but pumpsets use either electricity or diesel power.

In the cities the hot summer (reaching 45 °C by day and staying above 30 °C by night) is insufferable, unless a person can work indoors in an air-conditioned space, and sleep in an air-conditioned room. The power demand peaks in the summer, at the same time that the farmers' demands for pumping power are at their maximum. The government has so far, though, paid even less attention to the standards of building insulation to keep heat out than the British have, despite their poor track record, to keep heat in. The demands for power increase annually at a staggering rate, and the policy is to try to satisfy those demands from big dams and from thermal power stations burning India's large supplies of rather dirty coal, rather than by stemming demand. (It is perhaps not for Westerners to criticise the government here. Was demand management part of Europe's 'development'?) The combination of power demands by Bombay and of irrigation and drinking water demands by Gujarat are the essential driving force behind the contentious Narmada Valley Project, which has become a *cause célèbre* around the world because of its impact on some of India's remaining forested areas with both a rich wildlife and an indigenous tribal population.

RATES OF SOCIAL CHANGE

There is no objective way in which we can select particular indicators of change in society and say that they reflect the most important and fundamental changes going on. To some specialists nutritional indicators may be most important, to others the rate of accumulation of capital. Nevertheless we will start here by considering two indicators that seem fairly fundamental: the rate of population growth and the level and rate of urbanisation. It should not be forgotten that the most urbanised societies, which are also the most 'developed', have the lowest population growth rates.

Lewis (1978) puts this in historical perspective. Germany at the end of the nineteenth century was 50 per cent urban – and its population was growing at the rate of 1.2 per cent per annum. To accommodate this increase of population in urban areas it would therefore mean that the urban areas had to grow by 2.5 per cent per year – which was what they did. In Latin America the population was in 1978 also about half urban, growing at nearly 3 per cent per annum, meaning that for all natural increase to be accommodated in urban areas they had to increase by about 6 per cent per annum, which they just about succeeded in doing. But urbanisation is expensive – particularly in the infrastructure rather than just the productive plant required to support the population. Lewis (1978, p. 39) notes that the difference between lending and borrowing countries in the nineteenth century revolved around the rate of growth of their urban populations: 'Those whose populations grew by less than 3 per cent per annum (France 1.0, England 1.8 and Germany 2.5) loaned, and those whose urban populations were growing by more than 3 per cent

Table 11.2 Urbanisation in India 1901–91

Date	Total population (millions)	Rural population (millions)	Urban population (millions)	Annual growth rate for the decade	Rural annual growth rate for the decade	Urban annual growth rate for the decade	% urban	Migration as % of urban increase	Natural increase as % of urban increase
1901	238.5	212.6	25.9				10.9		
1911	252.1	226.2	25.9	0.56	0.6	0.0	10.3		
1921	251.3	223.2	28.1	−0.03	−0.1	0.8	11.2		
1931	278.9	245.5	33.4	1.05	1.0	1.7	12.0		
1941	318.7	274.5	44.2	1.34	1.1	2.8	13.9		
1951	361.0	298.6	62.4	1.25	0.8	3.5	17.3		
1961	439.3	360.3	79.0	1.98	1.9	2.4	18.0		
1971	548.1	439.0	109.1	2.24	2.0	3.3	19.9	20.9	62.2
1981	683.4	523.9	159.5	2.23	1.8	3.9	23.3	19.6	46.1
1991	846.3	628.7	217.6	2.16	1.8	3.2	25.7	21.7	58.0

Source: Cols 1–8, Census of India 1991; Cols 9–10, Visaria and Visaria 1994

per annum (Australia 3.5, United States 3.7, Canada 3.9, Argentina 5.3) borrowed.'

Table 11.2 shows several aspects at a decadal interval of India's population growth rates and urbanisation rates. It reveals a picture unlike Europe in the nineteenth century, or Latin America in the past few decades. As Lewis noted, if a country has a total population growth rate of about 2.5 per cent, and an urbanisation level of 20 per cent, then for all the natural increase to be accommodated in towns would require that the urban population grow by 12.5 per cent per year, which is 'virtually impossible'. In India's case, for the past four decades the total population growth rate has been at or above 2 per cent, and with an urban population of only about 18–19 per cent in 1961–71. The urban population has been growing at over 3 per cent for most of this period, but obviously nowhere near fast enough to accommodate all natural increase. The corollary of this can be seen in the figures for migration and local increase as components of urban growth – the last columns in Table 11.2. Most urban growth is attributable to the natural increase in the urban area. Urban areas are not actually absorbing that many rural migrants. (In absolute terms the figures are large, but not relative to India's overall population.) What of the future? I have made some projections for the Ganges Valley (Chapman 1995), where the majority of India's population live, and I have come to the conclusion that it seems likely that by 2030 the population will be only 40 per cent urban, although that percentage will still imply an increase in urban population by a factor of 4.7. The implications in terms of demands for water supply through the long dry season are awesome. As for the overall population of India, a recent paper by Seal and Talwar (1994) suggests the stabilisation of the population at around 1,480 million in

232

the year 2080 – let us say in crude terms a hundred years from now.

The overall conclusions of this cameo of Indian demographics are that both rural and urban populations are growing either fast or quite fast, that rural to urban migration does not account for much of urban increase, and that the level of urbanisation in India is increasing slowly by comparison with historical circumstances in Europe. Nevertheless, in absolute terms the urban population is growing very fast. In the past decade the increase in urban population was nearly 60 million people – around the total population of the UK added in just ten years to urban areas. These 60 million people are in search of housing, jobs, food, and education. Thirty to forty per cent of them live in squatter settlements and slums, without formal sanitation. An annual growth rate of 3.9 per cent implies a compounded decadal growth rate of 47 per cent, and one of 3.2 per cent implies 37 per cent. Individual cities of course score above-average increases – and it is particularly the big metropolitan cities that have recorded growth rates of above 50 per cent in a decade.

In India rural population densities have reached four times the peak density reached in Europe before industrialisation and urbanisation (and in Bangladesh eight times the density). India is thus in a particularly difficult predicament. Because towns cannot grow fast enough to accommodate all natural increase, rural areas will have to continue to increase both employment and food output – although there is no new land for the extension of cultivation – and at the same time India must face the infrastructural costs of urbanisation on a large scale. The country must also attempt to increase the rate of urbanisation wherever possible, to relieve pressure for job creation in rural areas, and to induce the changes in lifestyle and education which accompany urbanisation, and which result in lowered rates of natural increase. If necessary India must, and actually will, go fairly heavily into debt.

I have left to last what is perhaps the most important indicator of all. At present 40 per cent of the population live below the official poverty line – and the official line is set at a very low level. This poverty is manifest in inadequate employment, inadequate access to resources for self-employment, inadequate income, food intake, and housing. Despite the fact that 30–40 per cent of the urban population live in slums, urban areas have greater wealth, higher average income levels, better nutrition, higher levels of literacy and better access to health, family planning, and education, lower mortality rates (particularly among the young), and lower birth rates. Although there are exceptional states such as Punjab and Haryana, for most of India modernity and 'development' are more urban than rural.

ENVIRONMENTAL IMPLICATIONS

The Brundtland (World Commission on Environment and Development) definition of sustainable development is 'development that meets the needs of the present without compromising the ability of future generations to meet their own needs'. The problems of operationalising this definition are immediately obvious. It assumes without proof that the twin goals are reconcilable. It assumes that there is some absolute definition of need. It also makes a dubious distinction between the present and the future. The idea of a 'generation' dissolves when one thinks of the births and deaths occurring every day. Those currently alive will have legitimate demands in their lifetime, which extends up to seventy years into the future, for the relief of poverty, and by admitting now that their needs should be met we accept a future liability that may compromise future generations. And this argument can be repeated tomorrow, and the day after, and so on. Karshenas (1994, p. 734) discusses the definition give by Pearce et al. (1988, p. 6), which stresses not the human demands but the constancy of the natural capital stock; that is, non-negative changes in the stock and quality of natural resources such as soil, biomass, water. As Karshenas points out, the definition means that any degradation (a loss of soil to a brick plant for example?) is immediately taken as a sign of unsustainability. And further, no idea is given of the stock levels to aim for. Karshenas also develops an argument to suggest that to the extent that such a vague definition can be applied at all, it applies best to developed economies well within the sustainability frontier.

Karshenas therefore develops a new definition:

> Sustainable development may be defined in terms of the pattern of structural change in natural and man-made capital stock (including human capital and technological capabilities), which ensures the feasibility of at least a minimum socially desired rate of growth in the long run.

> (Karshenas 1994, p. 734)

He points out that many developing countries are near the sustainability frontier, not comfortably above it, and that below the minimum rate of growth such economies may face forced environmental degradation through poverty and lack of income-earning opportunities. In such circumstances 'it may be efficient for a developing economy, in transition from a natural resource based to an industrial economy, initially to run down its natural resource base in order to accumulate man-made capital'. He also notes that the present increase in forest cover in industrial countries is simply a recovery from their earlier depletion during their development stage. And, to return to the brick pit cited above, the use of brick and concrete in housing construction in rural areas in the Himalayas as a result of quarrying the source materials and building roads to connect rural areas results in a much lower use of timber in the building stock, and more durable buildings.

Karshenas concludes his paper with these words about a hypothetical developing country:

> Inadequate man-made capital, stagnant technology, lack of employment opportunities and inability to cater for basic human needs, combined with a growing population, have forced the economy into a state where survival necessitates eating into the natural or environmental capital stock. Project aid which is solely directed towards repairing the environment or replenishing the natural capital stock while disregarding the broader developmental issues, may turn out to be ineffective in such circumstances. Rebuilding of the stock of man-made capital and employment generation even in areas or sectors far removed from the immediate environment, may be a more effective means of environmental preservation. Under certain circumstances it may even be necessary to run down the natural resource stock in the short term in order to achieve these development goals. This is not, of course, to underestimate the significance of environmental policy as such, but to emphasise the broader developmental aspects of environmental policy which could be easily neglected if one takes a purely ecological perspective.
>
> (Karshenas 1994, p. 754)

This statement is eminently reasonable, but does of course conceal as many problematical elements as the definitions of sustainable development considered before. Environmental history, like human history, can never be repeated. The industrial nations may well be reafforesting their countryside, but they are not re-creating the original species mix of the original ecosystems. What therefore is a permissible as opposed to a non-permissible run-down of natural capital stock?

As pointed out above, independent India's quest has been the pursuit of 'development', whether as in the first four decades by state planning in a mixed economy, or, more recently, by more free-market approaches. The public investment in heavy industry and associated developments in the private sector turned India into a major industrial power, and helped to some extent to increase the level of urbanisation. But it did not increase non-farm employment by much at all, and direct employment in industry has stayed resolutely stuck at about 10 per cent. Employment growth in urban areas has been much greater in service sectors than at a comparable stage of development in the North. Both urbanisation itself and the growing industrial base have made major resource demands on the environment – in terms of demands for power, water, and raw materials. Where the raw material consists of forest products, the state has followed a policy which has allowed commercial exploitation at a large scale.

For the first two decades agricultural output did not increase as fast as population. In the mid-1960s successive monsoon failures caused what is now acknowledged to have been a famine resulting in widespread suffering and death, and resulting in a growing dependence on imported food. This was the first major 'failure' of independent India, and it has stuck in the public and political mind alike. In research into the media (see the next section) in India, most producers and editors spontaneously identified this period as the beginning of environmental awareness in India on a large scale. The awareness is therefore rooted not in international conferences nor global climatic change, but in failed monsoons. The Indianness of this experience was stressed by Darrel D'Monte, formerly editor of the *Times of India*, a committed environmentalist and president of the new International Association of Environmental Journalists:

> Climatic change ... this is not a problem India is going to worry about, because we have already experienced that. We've been suffering droughts for a long time. We've been suffering famines. We know what change is all about. Nobody can threaten us with any greater change than we've already had.

The blame for what had happened was in part put on deforestation – and reforestation has been taken seriously ever since.

But the response was also the adoption of green revolution techniques in agriculture including the use of new hybrid seeds, pesticides, fertilisers, and a massive increase in irrigation from both dams and canals and also from electrically powered tubewells. This has resulted more recently in food output increasing faster than the population, so that buffer stocks are now available at a national level to tide the country over one or even two bad monsoons.

Thus it is easy to detect the environmental impacts of both urban and rural sectors, at many scales and in many areas. An increasing concern for the 'degraded' state of the environment has fuelled a growing environmentalist lobby, spearheaded by well-funded and competent NGOs which have developed a world-wide reputation and world-wide contacts. The names of some of these, such as Sunderlal Bahuguna's Andalo Chipko movement for saving trees in the hills, have become icons in international green circles. In urban areas green lawyers skilfully use India's public interest legislation to pressurise industrial polluters – of whom there are many, discharging untreated wastes into watercourses and simply on to waste ground, whence it contaminates ground water. The worst urban pollution disaster, the killing of several thousand people by a gas leak from an agrochemical factory in Bhopal in Madhya Pradesh in 1985, has become synonymous, together with Chernobyl, with technological evil and disregard for human well-being.

In India, in other words, the die is well cast for a dispute between the environmentalist and the development lobbies. The kinds of issue which

Karshenas hints at are in the open here and debated regularly in the press. Is this 'development' really 'unsustainable', or is there a necessary reduction in natural capital stock in order to accelerate development to the point of better sustainability?

THE MASS MEDIA AND PUBLIC OPINION IN INDIA

Both John Gordon and Ros Taplin in their chapters in this book stress the significance of public opinion and its relationship with the political process. In the end it is not the scientists nor even the politicians as decision-makers that matter most – no matter what international *melas* (fairs) there may be. Patterns of consumption are not dictated by the politicians – even if their actions may effect the productiveness of the economy. In India, the world's largest democracy, public opinion matters in evaluating the environmental costs and the developmental gains of different strategies and policies. The last substantive part of this chapter looks at public opinion in India on these issues as a result of original research work carried out in the period 1992–4.

The first and most important thing to realise about communication in India is the extent to which there is not one but a multiplicity of publics. India's people, who number approximately 900 million, inhabit a tower of Babel. Depending on the definition used and the extent to which a dialect becomes another language, there are between 225 (1931 Census) and 1,652 (1961 Census) languages spoken in India, but obviously not all can have equal official status. Fifteen of the languages are Scheduled, that is to say recognised for official use in education and the law courts, etc. These fifteen have different scripts, and although for some of the northern languages derived from Sanskrit the scripts are very similar to the Devnagri used by Hindi, there are at least ten completely different calligraphic systems. Officially Hindi is the sole lingua franca, but attempts in the past to impose it in preference to English have met with stiff resistance, particularly from the south of the country. Thus English, which is not a scheduled language, is nevertheless used and accepted as a lingua franca in higher education, in business, and in government. The number of people who speak English as a mother tongue is negligible, but the number of people who learn to speak it with some degree of competence is of course significantly higher. Ancient India was noted for its scientific achievements, and in particular its contribution to mathematics, but anecdotal evidence suggests that contemporary Indian languages have poorer scientific vocabularies than English. In terms of scientific publication in India, English is overwhelmingly preferred.

Table 11.3 shows the distribution of newspaper publishing by language and state. Of course newspapers reach directly only the literate population. Table 11.4 shows the 1991 breakdown of literacy.

For most of the existence of independent India, radio and TV have been

Table 11.3 Number of daily newspapers in India by state of publication and language, 1993

Language \ State	Andhra Pradesh	Assam	Bihar	Goa	Gujarat	Haryana	Himachal Pradesh	Jammu and Kashmir	Karnataka	Kerala	Madhya Pradesh	Maharashtra	Manipur	Meghalaya	Orissa	Punjab	Rajasthan	Tamil Nadu
Assamese		5																
Bengali		3																
English	1	4	3	2	1			2	2		2	13	1	2	1		1	2
Gujarati					18							6						
Hindi		2	17			2	1				30	7			1	3	23	
Kannada									9									
Konkani				1														
Malayalam										13								
Marathi				2					2			33						
Nepali																		
Oriya															9			
Punjabi																12		
Sindhi												1					1	
Tamil									1									9
Telugu	8																	
Urdu	3							1	1		1	2						
Total	12	14	20	5	19	2	1	3	15	13	33	62	1	2	11	15	25	11

Table 11.4 Percentage literacy in India, 1991

	Total	Male	Female
All India	52.2	64.1	39.3
Rural	44.7	57.9	30.1
Urban	73.1	81.1	64.1

Source: Census of India, 1991

state monopolies, and partly used to spread the message of 'development'. At Independence in 1947 All-India-Radio (AIR) had only six stations and eighteen transmitters – covering 11 per cent of the population. By 1985 AIR, directly under the authority of the Ministry of Information and Broadcasting, had eighty-eight stations and 167 transmitters, covering 90 per cent of the population. By now transmission coverage should be nearly complete, but listening figures are of course lower: 64 per cent in urban and 46 per cent in rural areas in 1989. Much of the programme output is music, poetry reading, and drama, the latter in regional languages. All regions carry the news in English, Hindi, and the regional language. There are also documentaries, some associated with scientific developments, but anecdotal evidence (e.g. Malik 1989) always seems to suggest that popular music is what most listeners want from radio.

Television in India started in 1959 with help from UNESCO, very much with an eye to educating and improving the audience. A single station in Delhi broadcast to 180 teleclubs, which had been given free receivers. This was an experiment to see what TV could achieve in community development and formal education (Kumar 1994, p. 154). In 1961 educational TV programmes on science for schoolteachers started – and only in 1965 did the first entertainment transmissions begin, under pressure from the manufacturers of receiving sets. In 1967 farmers' teleclubs were started in rural areas near Delhi, to spread word of new agricultural techniques (the green revolution). From these small beginnings TV developed into a much more general public service broadcasting agency, broadcasting programmes in all usual categories, including entertainment and film. The idea of community TVs was not abandoned, and headmen in an expanding number of villages have been given such sets for community purposes, although there are a competing number of privately owned sets which are easier to see if the audience number less than ten rather than more than forty. It is also now common for the family of a bridegroom to ask for a TV set as part of the dowry from his wife, so clearly the potential for viewing is going up. However, power shortages and power failures are endemic in India, even more so in rural than urban areas. Viewing is likely to be disrupted, which means that the attraction of 'the box' is commensurately reduced.

Doordarshan (also known as DD, the state TV company) also makes programmes for educational television, broadcasting in school hours, and concentrating on science. Part of the reasoning is that few schools have laboratories for practical science work, and that TV provides an alternative means of demonstration. Take-up has been poor, in part for language reasons: narrators may use vernacular languages but not local dialects. Partly it is organisational – since education is a local state government subject, whereas TV is still centrally run.

Since 1984, in another development, the University Grants Commission Higher Education Project launched a Countrywide Classroom scheme, beaming educational programmes, part domestic, part foreign, to localised audiences. These have a viewership of 19 million (Kumar 1994) with a majority of watchers non-students. Overall the audience is still more urban than rural, and with 34 million sets in 1992 for a population of 860 million reach of TV is still limited, possibly to less than 25 per cent overall.

The latest development is the establishment of commercial channels (principally Zee TV) in the major cities, a second channel for Doordarshan in major cities, and of course satellite competition from Star TV (Agarwal 1994). The latter has only a small part of the audience; its programming in English with non-Indian accents must restrict its viewership, in addition to the obvious cost problems of dishes, or local (illegal but frequent) cable distribution. But it takes a very high share of advertising – a testimony to the élite nature of the small audience it finds. There is no doubt that this competition has shaken Doordarshan to improve on its drab and fusty image. The government is less inclined to intervene in the new era of liberalisation, and a new director-general is pushing his staff to modernise.

The transformation of broadcasting, from a serious but limited attempt at 'improvement' as part of a government-led 'development crusade' with a 'scientific temper', to a multifarious enterprise mostly sponsored by advertising reflects the consumer behaviour and consumer power of the emergent urban middle classes of industrialising and liberalising India. It is also consistent with the world-wide philosophy of the late 1980s, of privatisation, government retrenchment, and trade liberalisation. To many Indian critics all this represents the triumph of Western-based materialism, at considerable environmental cost, and the final abandonment of any Gandhian dream of rural development based on appropriate technology. But no one has yet suggested seriously a national Radio Gandhi.

The empirical data presented here come from an ESRC-sponsored research project into the mass media and environmental knowledge in India and the UK which looked at media content, media production, public knowledge, and public responses to environmental messages. The data presented here are taken from just the public opinion/knowledge survey in India alone, with some additional comments from interview material with Indian media producers. The questionnaire survey was conducted with 545 people in and around Pune in Maharashtra in western India, and 716 people in and around Coimbatore in Tamil Nadu in the South. The first sample is more urban and prosperous: even the rural people are influenced by the dynamic economy of Pune. The southern sample is comes from a slower-growing and more rural area – though one with a greater tradition of using English.

Respondents were asked what are the most important issues facing the world today (Table 11.5), the most important issues in India, and the most important issues locally. The answers were open-ended. The answers were classified by the author, using cover-set categories – that is to say, an answer could belong in more than one category (and therefore the percentages can add up to more than 100 per cent). Once an answer has been classified, any repeat use of that answer by another informant was automatically classified the same way by a computer program.

In the South violence (including terrorism) and poverty top the list, followed by communalism, war, and the environment. When these data are broken down by type of public, here defined by newspapers readership (English language, vernacular language, none), two things stand out immediately. The first is that the English-language audience give violence a much higher rating, and they put the environment second. For the other two groups poverty and economic issues are more important than the environment. We have established in our research that the English-language media (particularly the newspapers, but also TV in certain ways) tend to take a pro-environment stance, while the vernacular papers take a pro-development stand. On an issue such as the Narmada dam this is particularly clear. In the Pune sample, which includes no people in our low-income group, the environment does come out as the leading world issue overall, but again the breakdown shows some interesting variations. In the English audience war (a category including issues such as nuclear proliferation, not just active combat) and economic issues are as important as the environment; those who read the vernacular press actually put environment first, followed by population (pressure) and economic issues; whilst those who read no papers have virtually no opinions on any world issues. The English audience also put Northern hegemony high on their list of world problems.

At the national scale, in the South economic issues, population, and communalism dominate, and in Pune economic issues and population dominate. The environment is not seen as a major concern for India overall by either sample.

Finally we turn to the local scale. At this level it becomes increasingly difficult to aggregate the answers into environmentally related categories. There is a lot of concern over transport. The greatest concern is that there is not enough of it, nor enough roads for the traffic – that communication is in other words poor. There is also concern over the pollution that transport causes, and both views are offered simultaneously, but the demand for more transport is greater than the demand for less pollution. In the South

Table 11.5 Results of survey of knowledge of world and local issues in two parts of India

What do you think are the most important issues affecting the world today?

| | SOUTH | | | | | | WEST | | | | | |
| | | | | Newspaper | | | | | | Newspaper | | |
	Total	TV	No TV	English	Vernacular	None	Total	TV	No TV	English	Vernacular	None
	716	450	266	148	172	396	545	407	138	127	281	137
Don't know	40	28	60	9	29	56	27	9	80	1	10	88
Communalism	6	8	5	9	11	3	11	14	1	24	10	0
Corruption	2	2	2	3	2	1	5	7	0	3	8	1
Crime	2	2	0	3	2	1	0	0	0	1	0	0
Disasters	1	1	1	1	1	1	0	0	0	0	0	0
Economy	6	6	6	11	8	4	23	29	7	33	29	3
Environment	7	10	3	20	4	3	32	40	7	34	44	4
Health	6	9	2	11	8	4	3	3	1	6	2	0
Northern hegemony	2	2	0	4	1	1	7	9	2	17	6	0
Politics	3	4	1	5	2	2	2	2	1	4	2	0
Population	5	4	7	9	6	3	21	27	6	17	31	5
Poverty	9	10	6	8	7	10	7	8	6	11	8	2
Violence	17	20	12	30	20	12	10	14	1	13	14	1
War	6	9	3	10	9	4	16	21	4	35	15	1
Other	3	4	2	5	2	3	3	4	1	8	2	0

What do you think are the most important issues affecting India today?

Newspaper

	Total	TV	No TV	English	Vernacular	None
	716	450	266	148	172	396
Don't know	31	19	48	3	20	45
Communalism	19	24	11	35	22	13
Corruption	3	4	2	10	1	2
Disasters	0	0	0	1	1	0
Economy	18	21	14	28	26	11
Education	2	2	1	3	2	1
Environment	2	2	1	6	1	1
Health	2	3	0	3	1	2
Northern hegemony	0	0	0	1	0	0
Politics	11	13	9	18	13	8
Population	12	14	9	21	13	9
Poverty	9	10	9	10	10	9
Violence	8	8	7	10	8	7
War	0	0	0	0	0	1
Other	3	3	2	5	1	2

Newspaper

	Total	TV	No TV	English	Vernacular	None
	545	407	138	127	281	137
Don't know	8	1	27	1	0	30
Communalism	19	21	12	28	20	8
Corruption	12	15	4	17	14	4
Disasters						
Economy	46	54	24	46	60	18
Education	13	14	8	17	14	7
Environment	15	19	5	13	23	3
Health	2	1	2	2	1	1
Northern hegemony	10	13	3	20	11	1
Politics	9	11	1	20	7	0
Population	63	67	49	55	72	50
Poverty	12	14	7	18	12	6
Violence	3	4	0	6	3	0
War	0	0	0	0	1	0
Other	5	6	1	11	4	0

Note: The figures at the tops of columns are the sample size; other figures are percentages of sample size

Table 11.5 Continued

What do you think are the most important issues affecting your town or village today?

	SOUTH			Newspaper			WEST			Newspaper		
	Total	TV	No TV	English	Vernacular	None	Total	TV	No TV	English	Vernacular	None
Don't know	716	450	266	148	172	396	545	407	138	127	281	137
Communalism	17	14	24	5	13	24	4	3	6	8	2	3
Corruption	8	7	9	5	7	9	1	1	1	2	1	1
Crime	1	2	0	3	1	1	2	2	1	5	1	1
Disorganisation	0	0	0	1	0	0	0	0	0	0	0	0
Drinking water	9	8	11	10	10	8	9	11	4	24	6	2
Economy	14	14	15	14	13	15	27	27	28	6	35	30
Education	4	4	3	5	6	3	14	10	26	6	11	27
Electricity	4	5	3	5	3	4	4	4	3	3	5	1
Environment	9	12	5	19	12	5	2	3	0	0	4	2
Health	14	17	9	24	16	10	27	32	12	29	33	13
Northern hegemony							13	14	7	20	12	7
Politics	2	3	0	5	3	0	0	0	0	1	0	0
Pollution	2	3	0	7	1	1	1	1	1	0	2	0
Population	2	2	2	4	1	2	10	13	4	19	10	4
Poverty	6	5	9	1	4	9	19	17	27	24	14	26
Transport	19	19	19	20	24	16	10	6	21	5	6	21
Violence	1	1	0	1	2	0	37	40	26	43	40	24
Other	7	8	6	10	4	7	0	0	0	0	1	0
							2	2	1	6	1	0

Give examples of Indian policy which will improve the global environment

				Newspaper		
	Total	TV	No TV	English	Vernacular	None
Don't know	716 / 90	450 / 86	266 / 96	148 / 62	172 / 95	396 / 98
Appropriate technology	0	0	0	0	1	0
Conservation	0	0	0	1	1	0
Education	0	0	0	1	0	0
Forest	3	5	1	14	1	1
Health	0	0	0	1	0	0
International politics	4	5	2	14	2	1
Nothing	2	2	1	7	1	0
Pollution control	2	3	0	7	1	0
Population policy	0	0	0	1	0	1
Other	2	2	1	7	1	0

				Newspaper		
	Total	TV	No TV	English	Vernacular	None
Don't know	545 / 65	407 / 56	138 / 92	127 / 51	281 / 57	137 / 97
Appropriate technology	0	0	0	0	0	0
Conservation	2	2	0	5	1	0
Education	0	0	0	1	0	0
Forest	18	22	6	6	30	3
Health	0	0	0	0	0	0
International politics	9	12	0	23	6	0
Nothing	2	3	1	10	0	0
Pollution control	5	6	0	6	6	0
Population policy	0	0	0	1	0	0
Other	2	3	0	4	2	0

Note: The figures at the tops of columns are the sample size; other figures are percentages of sample size

Table 11.5 Continued

	SOUTH						WEST					
				Newspaper						Newspaper		
Give examples of Indian policy which will improve the national environment	Total	TV	No TV	English	Vernacular	None	Total	TV	No TV	English	Vernacular	None
Don't know	716	450	266	148	172	396	545	407	138	127	281	137
	82	78	93	53	81	94	56	45	90	44	42	93
App. technology	0	0	0	1	0	0	1	1	0	1	1	0
Biodiversity	1	1	0	3	0	0						
Conservation	1	2	0	5	1	0	3	4	0	6	2	0
Economy	2	3	0	5	2	0	1	1	0	4	0	0
Education	0	0	0	0	1	0	1	1	0	1	1	0
Forestry	10	14	4	22	15	4	27	34	7	10	44	6
Intpol	0	0	0	0	0	0	0	0	0	1	0	0
Nothing	0	0	0	1	0	0	2	3	0	8	0	0
Pollution control	1	2	0	5	1	0	10	13	2	26	7	1
Politics	1	1	0	1	1	1	0	0	1	2	0	0
Population	1	1	0	3	1	0	0	0	0	0	0	0
Poverty	0	1	0	0	0	1	5	6	1	6	6	0
Other	2	2	1	7	1	0	1	1	1	4	0	0

Give examples of Indian policy which will improve the local environment

Newspaper

	Total	TV	No TV	English	Vernacular	None
	716	450	266	148	172	396
Don't know	78	72	89	53	79	88
App. technology	1	1	0	1	1	0
Conservation	2	2	0	3	2	1
Economy	0	0	0	0	1	0
Education	5	8	0	13	5	3
Afforestation	0	1	0	1	1	0
Garbage	2	3	0	5	3	0
Health						
Nothing	7	6	7	8	8	6
Pollution control	1	1	0	3	0	0
Population	1	1	0	1	1	0
Transport	0	1	0	1	1	0
Water conservation	1	1	0	3	0	0
Other	3	3	2	7	1	1

Newspaper

	Total	TV	No TV	English	Vernacular	None
	545	407	138	127	281	137
Don't know	52	41	87	41	38	92
App. technology	1	1	0	3	0	0
Conservation	1	1	0	2	1	0
Economy	1	1	0	1	1	0
Education	1	1	0	0	2	0
Afforestation	28	35	8	12	46	7
Garbage	1	1	1	3	1	0
Health						
Nothing	3	3	1	10	0	0
Pollution control	6	8	1	16	5	1
Population	1	1	0	0	1	0
Transport	1	1	0	2	0	0
Water conservation	8	9	2	10	9	1
Other	1	1	1	3	1	0

Note: The figures at the tops of columns are the sample size; other figures are percentages of sample size

Table 11.6 Results of survey of knowledge of global warming and ozone holes in two parts of India

SOUTH

Have you heard of global warming?

	Total	TV	No TV	Newspaper English	Newspaper Vernacular	Newspaper None
	716	450	266	148	172	396
No response	8	6	12	6	11	7
No	79	77	83	53	79	90
Yes	13	17	5	41	10	3

What will it do?

	Total	TV	No TV	Newspaper English	Newspaper Vernacular	Newspaper None
	716	450	266	148	172	396
Don't know	80	85	96	64	91	98
Climate change	7	11	2	27	6	1
Disasters	0	0	0	1	0	0
Economic harm	1	1	0	0	2	1
Env. harm	1	1	1	1	3	0
Flood	4	5	2	13	3	1
Harm health	1	2	0	4	1	0
Other	1	1	1	3	0	1

WEST

Have you heard of global warming?

	Total	TV	No TV	Newspaper English	Newspaper Vernacular	Newspaper None
	545	407	138	127	281	137
No response	6	6	4	1	8	4
No	46	32	87	9	40	93
Yes	48	62	9	90	52	3

What will it do?

	Total	TV	No TV	Newspaper English	Newspaper Vernacular	Newspaper None
	545	407	138	127	281	137
Don't know	54	39	94	11	51	98
Climate change	37	48	4	73	37	1
Disasters	0	0	0	1	0	0
Economic harm	0	0	0	0	0	0
Env. harm	5	6	1	9	5	0
Flood	13	16	4	31	10	1
Harm health	3	3	1	2	4	1
Other	1	1	0	1	1	0

What causes it?

	Total	TV	No TV	Newspaper		
				English	Vernacular	None
	716	450	266	148	172	396
Don't know	93	89	99	74	94	99
Deforestation	0	0	0	1	1	0
Energy consumption	1	1	0	2	0	0
Env. harm						
Greenhouse gas	4	6	0	14	3	1
Ozone	1	1	1	3	1	0
Pollution	1	2	0	5	0	0
Population						
Other	1	1	0	3	1	0

	Total	TV	No TV	Newspaper		
				English	Vernacular	None
	545	407	138	127	281	137
Don't know	57	46	94	20	54	98
Deforestation	16	21	1	34	15	1
Energy consumption	1	1	0	3	0	0
Env. harm	3	4	0	4	5	0
Greenhouse gas	3	3	1	8	2	0
Ozone	10	12	2	17	10	1
Pollution	23	30	2	44	24	1
Population	0	0	0	0	0	0
Other	1	2	0	2	2	0

Have you heard of the ozone hole?

	Total	TV	No TV	Newspaper		
				English	Vernacular	None
	716	450	266	148	172	396
No	80	73	92	40	81	95
Yes	20	27	8	60	19	5

	Total	TV	No TV	Newspaper		
				English	Vernacular	None
	545	407	138	127	281	137
No Response	3	3	4	2	4	5
No	43	29	86	4	37	91
Yes	53	68	9	94	59	4

Note: The figures at the tops of columns are the sample size; other figures are percentages of sample size

Table 11.6 Continued

SOUTH

Which hemisphere(s) is it/are they in?

	Total	TV	No TV	Newspaper English	Newspaper Vernacular	Newspaper None
	716	450	266	148	172	396
Don't know	86	81	94	57	85	97
Both	6	8	4	19	6	2
North	7	10	2	20	8	1
South	1	1	0	4	1	0

Why is it important?

	Total	TV	No TV	Newspaper English	Newspaper Vernacular	Newspaper None
	716	450	266	148	172	396
Don't know	82	82	95	56	89	97
Climate change	1	2	0	5	1	1
Env. harm	5	8	0	16	6	1
Health	4	5	2	17	1	1
Other	2	3	1	9	1	1

WEST

Which hemisphere(s) is it/are they in?

	Total	TV	No TV	Newspaper English	Newspaper Vernacular	Newspaper None
	545	407	138	127	281	137
Don't know	51	37	93	14	45	97
Both	16	20	3	32	16	1
North	27	35	3	43	31	2
South	6	8	1	10	8	0

Why is it important?

	Total	TV	No TV	Newspaper English	Newspaper Vernacular	Newspaper None
	545	407	138	127	281	137
Don't know	53	41	91	16	48	97
Climate change	7	8	3	6	10	0
Env. harm	8	10	3	13	9	1
Health	27	35	3	52	28	3
Other	6	8	3	17	5	0

What causes it? (open question)

	Total	TV	No TV	English	Vernacular	None	Total	TV	No TV	English	Vernacular	None
				Newspaper						Newspaper		
				English	Vernacular	None				English	Vernacular	None
	716	450	266	148	172	396	545	407	138	127	281	137
Don't know	96	96	98	86	97	99	99	99	98	99	99	99
Pollution	2	2	2	6	1	0	0	0	1	0	0	1
Other	2	3	1	7	2	0	0	0	1	1	0	0

What causes it? (prompted)

	Total	TV	No TV	English	Vernacular	None	Total	TV	No TV	English	Vernacular	None
				Newspaper						Newspaper		
				English	Vernacular	None				English	Vernacular	None
	716	450	266	148	172	396	545	407	138	127	281	137
Don't know	87	81	97	57	89	97	50	36	91	9	45	97
Aerosols	5	8	0	16	6	1	10	13	1	14	12	1
CFC's	8	10	3	27	5	2	39	49	7	73	41	1

Note: The figures at the tops of columns are the sample size; other figures are percentages of sample size

transport, health, and economic issues dominate, and the pattern between different publics is not so great. In the West, transport, population, the provision of drinking water, and the environment dominate – though for the non-newspaper readers the environment is less important and poverty and economics more so. The environment category for this question covered a wide range: general complaints about industrialisation, lack of sanitation, lack of drainage, lack of general cleanliness, excessive garbage, deforestation, monkey infestations, and general degradation.

Questions were then asked about Indian environmental policy at the global, national and local level. In the South, there is virtually no idea of any Indian global policy, whereas in the West there is some response – namely that India's afforestation programme is partly a response to global issues. In the South, there is some idea of afforestation at a national level, but there is no idea of any local-level policy. In the West, at the national scale the don't knows dominate, but the English-language public cite pollution control as a national policy to some extent, and others cite afforestation. At the local scale it is all about afforestation.

Knowledge on specific global environmental issues was also tested (Table 11.6). In the South, awareness of the ozone hole(s) is poor, and over-whelmingly those who know about it are the English-language readers. In the West more people know about it, but those who know most about where it is and how and what causes it are again the English-language readers. A very similar set of findings comes from an analysis of the responses to questions on global warming.

In pursuing the language issues further we asked several producers and editors how they handled difficult topics such as greenhouse gases in vernacular languages. The answer was that such terms were very difficult to translate – that usually the vernacular alphabet was used to write the sound (phonetic value) of the English, followed by some local approximation in the local language.

In a second approach we asked students who were bilingual to translate particular words from English into a regional language, and then other students to translate back from the regional language to English. The word 'drought' mostly comes back as 'famine'. When Indian colleagues were asked about this, many had not understood there was any difference – they use the word 'famine' to mean 'drought'. This of course refers back to India's dependence on the monsoon, and the fact that famine is the result of a failed monsoon. 'Climatic change' comes back mostly as 'changes in the seasons' or 'change of weather'. This result I have again asked about, and there is an agreement that the very idea of climate – a long-term statistical average – is lacking.

The conclusions from this section are as follows. First, for the poor in India, who constitute the majority, the issues of poverty and lack of economic opportunity outweigh other issues, and to some extent the issue of communalism does too. For them the environment is important only at a local level, and then only in terms of what has not yet been improved by development; that is, the lack of sanitation or good water. Second, only for the English-educated élite is the environment important, but even then their knowledge of global issues is not strong. Finally, it is very difficult to translate many of the issues promoted by environmentalists in the North into local Indian languages.

The time-scales implications of this are, first, that for most people the national project to 'develop India' is the most important issue. It has widespread support and momentum, and is seen as a need. Helping to speed up the development of India will advance the time when the 'environment' in a Western sense will take higher priority. Second, ideas about the environment are bound up in language – and so far most of the debate at the international level is based in English, and reaches those who are literate in English. For these ideas to become more widespread the public in India needs to be more literate – again something which has moved slowly as India has developed slowly – and the local languages have to accommodate alien concepts in a meaningful way. This is clearly a long-term process, although I would not care to put a time bound on it.

What this section has not done for lack of space is to look in detail at how Northern and Indian media interact. The short answer is that they interact little – that the Northern groups are not particularly interested in reaching the Southern audience, least of all with environmental issues, and that the Indians are far more interested in themselves anyway. It is true that if they take international coverage of an issue, it is likely to come from a Northern agency, but such issues are in fact not frequently taken up.

CONCLUSIONS

The 'environment' is usually known in eponymous terms: the Russian environment, the Indian, or the Chinese. ˇ is recognised as such because human beings have divided the earth's surface into sovereign states, each with the right to manage its own affairs in its own way, such rights being surrendered voluntarily only where mutual cooperation seems beneficial, or forcibly where coercion is used to annexe or secede territory. Some states on this earth have long and stable histories; Switzerland is a prime example. But many do not, and when instability descends into conflict as in former Yugoslavia, the cost in terms of human life, economic capital, and environmental resources is horrendous. It is important therefore that the process of state or nation building, of the integration of the citizens of the state, be encouraged as a precondition of any policy in any other sphere. The integration within the state and the legitimisation of the state can then proceed only as far or as quickly as the state can help in providing its citizens with the basic necessities of life. The establishment of national consciousness and the development of the state go hand in hand. Both of these are long-term projects, but if they do not occur fast enough there is always the possibility of regression and what Karshenas termed forced environmental degradation. The faster that development occurs, the sooner this threat will recede, and it may well be necessary to run down some of the natural stock in the process.

In India's case the modernisation and improved demographic indicators that accompany urbanisation, together with the pressure of population in rural areas, mean that urbanisation, currently too slow, should be encouraged as much as possible. Because of the undeniable problems of the mega-cities, it is necessary to adopt policies to encourage the growth of medium-sized cities spread more widely through the country. But the servicing of these cities will never be easy, because of the extreme seasonality of the climate.

There is no reason why there should not be a dialogue between India and the outside world on ideas of environmental stewardship, but it is clear that there has to be patience with the linguistic and educational barriers, which can be reduced only slowly over the medium to long term. The fact that there can be easier communication with a small English-speaking élite should not lead to the misapprehension that one is communicating with the larger sections of Indian society – who are far more pro-development and far less pro-environment (in the Northern sense) than the élite, who constitute in a sense those who already 'have development'.

India is a democracy bound by electoral cycles of the sort that John Gordon has described, which leads to some degree of short-term thinking. But it is quite clear that in the case of India's attitude towards the environment the short-term electoral cycle is not actually of much significance, because the priority of 'development' has universal acceptance. Let me finally stress that Indians by and large have little knowledge of and little concern for global issues. Their subcontinent is large enough and complex enough to constitute for its citizens its own universe, and besides, few have the money or inclination to travel elsewhere. If the North would like India to think globally, then it will itself have to act globally by helping India in every way to reach acceptable levels of development for its citizens. When this long-term project has been achieved, when India can live more like the North, it may well act more like the North in other ways too.

REFERENCES

Chapman, G. 1993. Thinking about canal irrigation systems. Paper presented to Beijing Symposium on Irrigation and Drainage Systems, May 1993: Proceedings forthcoming, Wooldridge, R. (ed.) Hydraulics Research Ltd, Wallingford.
Chapman, G. 1995. The Ganges and Brahmaputra basins. In Chapman, G. P., and Thompson, M. (eds) *Water and the Quest for Sustainable Development in the Ganges Valley*. London: Mansell, 3–24.
Chapman, G. P., and Thompson, M. (eds) 1995. *Water and the Quest for Sustainable*

Development in the Ganges Valley. London: Mansell.

Karshenas, M. 1994. Environment, technology and employment: towards a new definition of sustainable development. *Development and Change* 25: 723–56.

Kumar, K. J. 1994. *Mass Communication: A Critical Analysis*. Bombay: Vipul Prakashan.

Lewis, W. A. 1978. *The Evolution of the International Economic Order*. Princeton, NJ: Princeton University Press.

Malik, S. 1989. Television and rural India. *Media, Culture and Society* (Sage, London and New Delhi) 11: 459–84.

Pearce, D., Barbier, E., and Markandya, A. 1988. Sustainable development and cost–benefit analysis. Paper presented at the Canadian Environment Assessment Workshop on Integrating Economic and Environmental Assessment; cited in Karshenas, M. 1994. Environment, technology and employment: towards a new development of sustainable development. *Development and Change* 25: 723–56.

Seal, K. C., and Talwar, P.P. 1994. The billion plus population: another dimension. *Economic and Political Weekly*, 3 September 1994, 2344–7.

Vasudeva, S., and Chakravarty, P. 1989. The epistemology of Indian mass communication research. *Media, Culture and Society* (Sage, London and New Delhi) 11: 415–33.

Visaria, P., and Visaria, L. 1994. Demographic transition: accelerating fertility decline in the 1980s. *Economic and Political Weekly*, 17–24 December, 3281–92.

12

CONCLUSION

Thackwray S. Driver and Graham P. Chapman

INTRODUCTORY COMMENTS

The previous chapters clearly show how different researchers or activists working in different disciplines have different time-scales in mind when considering environmental change; and they show how considerations of time can shed new light on oft-studied environmental issues. In this chapter we offer one possible framework to help understand the different time-scales used by different contributors, but both they and the reader may well wish to devise other frameworks too.

As we stated in the Preface, our intention has been to illuminate and contrast the way different academic disciplines conceptualised time and environmental change. Perhaps inevitably, given the subject matter, most chapters in this book take an interdisciplinary approach and it becomes hard to identify discrete disciplines in the traditional academic sense. On the plus side, throughout the book there are examples of how interdisciplinary approaches result in interesting and sometimes unexpected results. When a historian bothers to read ecology (Beinart, Chapter 8) he is able to question the tendency to conflate environmental change with destruction. Likewise when a physical geographer uses historical methods she is able to trace the detailed fluctuations in climate over a millennium (Grove, Chapter 3). Every contribution shows that there is much to be gained from explicitly considering the issue of time when examining environmental change; but is it possible to move beyond these individual studies and draw some more general conclusions, other than 'it is good to look at different time-scales'?

In what follows we discuss our temporal understanding of the environment in a kind of cross-tabulation; on one axis we have conceptualised time-scales of different magnitude – 'now time', the generational time-scale (10–100 years), the Century time-scale (100–1,000 years), and the Late Quaternary[1] time-scale (1,000–100,000 years) – and on the other axis we have the four explanatory paradigms of Table 11.1: the physical/chemical, the artificial, the biological, and the conscious–reflexive.

'NOW TIME'

We live only in the present; all else, the past and future, has to be imagined in some way. The past we can imagine at this 'now-time' through different memories: some individual, some collective in a society or culture, some interpreted from ever more carefully and widely collected data. We attempt to foresee the future, but with an increasing time horizon the uncertainties about the state of the future increase in tremendously to the point where it becomes rational to discount the future heavily in current decision-making.

At the current time, now, there seems to be a fairly widespread sense of urgency about environmental issues among many commentators. It may seem naive to ask why since that would obviously invite a heap of scorn and abuse from all sides, with comments such as 'none so blind as those who can't see'. But we are going to persist with this question for just a moment.

We live in the present; our understanding of our possible futures depends upon our understanding of both the past and how we came to be where we are 'now' and also of what we see to be the changes going on around us 'now.' Understanding of the past *may* enable us to extrapolate to the future and our understanding of the present trends *may* enable us to project into the future.

The problem is that both approaches are faulty. Consider first the idea of extrapolation from past trends and the coal question, which posed by a founding father of economics, W. Stanley Jevons, in 1865. He was clearly aware that Britain's international political power fundamentally rested on coal and that the coal reserves appeared limited so that by the middle of twentieth century Britain would be emasculated as the coal ran out. The idea that we would actually be shutting mines at the end of the twentieth century at the same time as we were discovering yet more seams of coal would have been fantastic: in later editions of the book he mentioned oil but dismissed it as some minor curiosity with a few limited uses. Extrapolation usually presumes too much about the stability of the underpinning variables to be of much use. It ignores the very complex feedback loops operating between resource exploitation, technological change, and social and political adaptation.

If, on the other hand, we actually know a current rate of change then we can project into the future. A current rate of change is essentially a first differential (in terms of calculus): the slope of some function at this point in time. But to be of use for predictive purposes the function has to be continuous – smooth (even if curving) in some way. Some functions obviously are, but modern thinking about chaos theory and fractals (fractals are self-similar at any scale of resolution, like the coastline of England) shows that many dynamic traces cannot in any sense be thought of as having any particular trend at any part. What slope is detected at any point is an arbitrary result of the scale of resolution, and has no predictive power. This is a problem every stock-market analyst would recognise. For us the problem is that we do not know what system may behave chaotically, or may change to behaving chaotically if (unknown) thresholds are exceeded.

To answer our original naive question, part of the reason why there is an increased sense of urgency 'now' is that there are some 'bad' trends which seem amenable to extrapolation. An obvious one is the growth of the human population and an obvious case study might be India. The 'horror' of population growth is normally illustrated by compound (exponential) growth: a line swinging ever more steeply upwards. It depends upon the fact that we can make a fairly accurate statement about current growth rates; for a nation state these are not arbitrary depending on the scale of resolution, since that is fixed at the national scale. On the ring road in South Delhi there is a billboard with an electronic display counting one by one as the number of Indians increase. At present about 45 children are born every minute, so the 'clock' adds 1 nearly every second – and it seems a fairly steady rate when you watch the clock. But the clock is speeding up. Currently it only takes 4.27 years to add 100 million people – but if the growth rate stays at 2.5 per cent, the time it takes to add another then another 100 million will fall rapidly (Table 12.1). The Indian government's belief is that the quickest way to bring the growth rate down is to achieve 'development' for its people as close to now as possible.

Another trend which seems amenable to extrapolation is the extinction rate of species. Hence the concern over loss of biodiversity, a diminution of natural capital stock which will deprive our successors for ever. The reason why this seems amenable to extrapolation is because we believe that we have a good-fit simple model: as human interference of the ecosystem increases, habitats are lost, food chains disrupted, and species stressed to extinction. No complex feedback loops are involved – so the possibility of a miss-forecast à la Jevons is assumed away. (We are aware that in prophetic terms species extinction may also be increased through anthropogenic climate change – see Wallis's chapter, Chapter 5.)

These two cases would appear to be grounds for current urgency 'now' – though given the low resource demands of the average Indian even that may be contested. But we are not sure that there are many trends which can be extrapolated or projected so easily. Beinart's and Fairhead and Leach's chapters (8 and 9) show how easy it is to mistake a 'current trend' and project

Table 12.1 Population and years taken to increase population by 100 million

Population (millions)	Years taken to increment population by 100m with growth rate at 2.5%
900–1000	4.27
1300–1400	3.00
1900–2000	2.08

to a dubious conclusion. Grove's chapter (3) shows how in some societies feedback loops enable adaptation and survival in the face of change, whereas in other societies the same flexible response does not operate. The success of adaptation depends on both the rate of environmental change being faced and the rate of change possible in the adaptive response.

If both rates are amenable to adjustment by human action then clearly there could be some attempts to find optimal balance between the two, by adoption both of mitigative measures with some attendant costs and of enhancing adaptive changes. If anthropogenic climatic change can be reduced by greenhouse gas emission abatement at some acceptable cost, and some smaller necessary adaptation of economics to changed climatic regimes is anticipated and supported, that might be a more sensible solution than a drastic precautionary scenario at one extreme, or a business-as-usual scenario at the other extreme.

The interesting point about climatic change is that we are not sure that we can predict the future by extrapolating from the past – because, as Clifford and McClatchey (Chapter 4) show, we have essentially a chaotic trace with a fractal self-similar quality about it. Similarly the changes in climate during the Quaternary, oscillating between ice ages and interglacial thermal periods, suggest that there are regions of phase space the system may inherit, just like the two lobes of the Lorenz attractor (Figure 1.1), but we do not know when the system may flip from one lobe to the other – and it may do so because of some sensitivity to a very small change in some initial conditions. Further, we do not know whether human activity will induce some 'flip' in the system. On the other hand we have the projections – based on the general circulation models – that underlie the thinking of the Intergovernmental Panel on Climatic Change. These are models which are often criticised for omitting some essential feedback loops, particularly loops involving the biosphere rather than just the physical and chemical properties of the atmosphere, the earth's surface, and the oceans (and even these are included only to a limited degree), and which avoid chaotic behaviour because that would make projection impossible.

There is one extrapolation which we can make with some degree of confidence: that the amount and quality of data available to modellers will increase, that their sophistication and complexity will increase, and that there will be fundamental conceptual changes in our modelling techniques. The result will be that we do not have a smooth trend to some more precise understanding of the world's climatic future: the predictions of futurology will themselves be liable to abrupt and discontinuous change.

We can sense directly the small scale in time and space, the here and now. It becomes progressively more difficult both to know and to understand as either or both the spatial and temporal scales expand. To some extent the history of humanity's understanding of the planet it inhabits is the expansion of its understanding of different temporal and spatial scales, of the new continents and new journeys to their interiors,[2] and of the discovery of the antiquity of earth and the evolution of life upon it. But broad temporal and spatial understanding is not commonplace; the bias in understanding is and will remain towards the here and now. This book has moved from the long time spans to shorter ones and the future. Here in the conclusions we push back from the here and now progressively again to the longer scales.

SUSTAINABLE DEVELOPMENT AND THE GENERATIONAL TIME-SCALE

development to be sustainable must meet the needs of the present without compromising the ability of future generations to meet their own needs.

(World Commission on Environment and Development 1987)

The Brundtland definition of sustainable development (p. 234), which has dominated discussions about the environment since the late 1980s, is based upon the idea of inter-generational equity. The basic unit of time in many discussions about the environment may, therefore, be said to be about twenty to thirty years – what we have termed here the generational time-scale. Of course, dividing passing time into generations is a strange abstraction: a new generation is born every day. Nevertheless, the concept of generations dominates debates about future environmental change.

The Brundtland definition does not make any suggestions about the number of 'future generations' we should take into account when considering our impact on the environment. As Wallis (Chapter 5) shows, most global circulation models – and the attendant suggestions for greenhouse gas policies – have been calculated with a cut-off time horizon of about five generations (i.e. 100–150 years). For any policy decisions there has to be either a discounting or cut-off time horizon of future environmental impacts of current activities; even ignoring the theoretical issues addressed by Faber and Proops (Chapter 10), any assessment of the needs of future generations can consider only a limited number of future generations.

The concept of inter-generational equity is based upon an obvious human emotion – the wish to protect our children and grandchildren; talk of 'future generations' rarely stretches beyond a few generational time-units. Proponents of sustainable development perhaps see in the concept 'future generation' a way of balancing the individual with the collective. Advances in medical science, improved nutrition, better sanitation, and clean water now mean that over much of the world people are living longer than their parents and grandparents. Many of us can now expect to live through two, three, or even four generational time-units. Ironically, many of us will be the very resource users in the 'future generations' that the Brundtland Commission urges us to consider. While the Brundtland definition may, at first sight, appear to demand a radical rethink in policymaking there are already some sectors of the economy that are used to thinking in similar time-scales – such as the insurance and pensions industry.

From our present vantage point at the end of the twentieth century we believe we are able to trace significant developments in the world of conscious human thought over a generational time-scale. Taplin (Chapter 6) and Gordon (Chapter 7) show how the adversarial political system that has developed in countries such as the UK and Australia hinders the enactment of environmental legislation. Their analysis also indicates that politicians' unwillingness to challenge the centrality of one concept, economic growth, because of a belief that this represents the 'will of the people', means that environmental policy will always be piecemeal. In the Indian context, Chapman (Chapter 11) shows how the concept of development also limits environmental policy – but he also shows that this particular ideology of development came about in the years immediately before and after Indian independence because the political élite needed one big idea around which they could forge a national identity.

How long and why do great ideas persist? Was the life span of communism an arbitrary period or does it relate in some way to the accumulated experiences and dialectical responses of only a few generations of people? The life span of communism is certainly related to the life span of the Cold War – a central state that oriented massive, if not nearly monopolising, scientific research expenditure along very specific military lines. It is not facile to suggest that the environment and development debate would not and could not have moved centre stage in international politics if the Cold War had not ended. In a similar vein, what will prolong or extinguish the idea of development in India? Will the Green movement there supersede current 'development' with a better idea?

In the same way there is a common perception that nature, the biological world, is unchanging except via human agency. 'Freak' events, such as hurricanes, may temporarily alter woodlands or coral reefs, but over time nature will re-establish itself – and all will appear as before. Ecology has traditionally added weight to this view: the biological world is described as a system with many negative feedback loops which maintains a dynamic equilibrium, with change taking place only at much longer, evolutionary time-scales. Without human interference the biological world would tend towards climax communities adapted by the physical/chemical world of climate and lithology. This view of ecology has now largely fallen out of favour. The new paradigm of ecology is based upon the recognition that change can on occasions take place at a much quicker rate. External shocks to the system, the 'freak' events, may be more important in affecting the characteristics of particular communities of plants and animals than any internal dynamics.

Both Fairhead and Leach (Chapter 9) and Beinart (Chapter 8) make mention of the growing literature on African environments (particularly dry grassland areas) that argues that the concept of disequilibrium is a better tool for understanding these ecosystems than is dynamic equilibrium (see, for example, Behnke *et al.* 1993). Nevertheless, both chapters emphasise the importance of human agency, in particular the changing political economy, in shaping today's environment. It is in the realm of the biological world that the majority of debates about environmental change are centred at a generational scale – particular when it comes to Northern beliefs about the environment in the South. As both chapters on vegetational change display (Fairhead and Leach, Chapter 9, and Beinart, Chapter 8), there is a strong assumption that all human impacts on the biological world are necessarily destructive. The subjective nature of many of these assumptions, usually based on Northern perceptions of what is a 'good' environment, needs to be carefully examined by the environmental movement.

Rain forests have been a central concern of the Northern environmental movement: the numerous books, TV programmes, and magazine articles concerned with their protection almost always make mention of their great age and stability before the arrival of 'Western man'. Their great age has become an article of faith among many environmentalists in the North, yet as Fairhead and Leach (Chapter 9) show, many areas now designated ancient rain forests are relatively young – sometimes as young as eighty years. Central to both Fairhead and Leach's and Beinart's chapters is a willingness to consult historical records beyond the one or two generations assumed to be sufficient by many social and environmental scientists interested in vegetational change. By stretching beyond the generational time-scale into the century time-scale they reverse some very widely held assumptions.

THE CENTURY TIME-SCALE

Above, we have argued that the concept 'future generations' used by many talking about sustainability in reality rarely translates into anything much above a century to a century and a half. But as Wallis (Chapter 5) illustrates, many of our present potentially environmentally damaging activities could have consequences over much longer time-scales. While much of the carbon dioxide (CO_2) being released at the moment will be absorbed into oceans within a few generations, a significant proportion will persist in the atmosphere for at least a few centuries. The rate at which CO_2 will dissipate is, however, a matter of great uncertainty; if global circulation models are bad at predicting climate change over a generational time-scale they are even worse over a century time-scale.

The major problem facing people trying to model future environmental change over a century time-scale is, of course, the physical world's own inherent instability. As Grove (Chapter 3) demonstrates, over these time-scales climate is in a constant state of flux and average global temperatures can vary significantly. Because of the problems of calculating present trends in climate change (Clifford and McClatchey, Chapter 4), we do not even know if the global circulation models are being calculated on a conveyor belt moving backwards or forwards.

What both Grove (Chapter 3) and Wallis (Chapter 5) show is that, for both the past and future, the century time-scale is vital to an understanding of debates about environmental change. Yet this time-scale receives scant regard in much of the literature on environmental issues; most texts on environmental change over periods longer than the generational time-scales described above are the work of physical geographers interested in late Quaternary, and longer, time-scales. If this book were to make just one strong recommendation it would be for increased research interest in environmental change over the century time-scale – especially in areas other than climate change.

It is over the century time-scale that humans have developed the industrial techniques that are now blamed for polluting the physical world and destroying much of the biological world. Support for the 'South' position on climate change (outlined in Chapman, Chapter 11 and the Editors' Note on Chapter 5) will be strengthened by calculations of atmospheric pollution that go back longer than a century to the birth of European industrialisation. It is over the century time-scale, too, that we see the rise and decline of European imperialism that has so transformed the biological world; from the deliberate plundering of Asian, African, and American resources to the unintended spread of new crops, weeds, and diseases (see Crosby 1986).[3] It is over the century time-scale that we can starkly expose the myth of stability in the biological world that drives our fears over rain forest loss at a generational time-scale.

In the realm of conscious human thought the century time-scale is one that, obviously, receives a great deal of attention. It is over this scale that we see the rise and fall of grandly titled intellectual eras; the Enlightenment, the Age of Reason, and so forth. From the perspective of the last decade of our millennium these shifts in the central organising principles of intellectual thought appear to be of great significance. Historians, literary theorists, and social scientists of all varieties can trace developments in the sphere of human activity against developments in the sphere of conscious human thought; the force of certainty behind European imperialism grew out of an ordered Newtonian universe and Darwinian[4] explanations of the natural domination of one race over another.

Fernández-Armesto (1995) has attempted to trace the development of some of these grand themes over the past millennium, but he does so from an interesting perspective:

I have a vision of some galactic museum of the distant future, in which

Diet-Coke cans will share with coats of chain mail a single small vitrine marked 'Planet Earth, 1000–2000, Christian Era'.... The distinctions apparent to us, as we look back on history of our thousand years from just inside it, will be obliterated by the perspective of long time and vast distance. Chronology will fuse, like crystal in a crucible, and our assumptions about the relative importance of events will be clouded or clarified by a terrible length of hindsight.

(Fernández-Armesto 1995, p. xiii)

Fernández-Armesto's brave attempt to write a world history from this perspective produces some interesting results.[5] European world hegemony no longer seems as all-powerful and monolithic; indeed, it is only truly apparent in the last quarter of the millennium, and even that hegemony is now cracking as the centre of world economic might shifts from the Atlantic world back to the Pacific.

Fernández-Armesto has used a century time-scale in order to move away from a Eurocentric history; social scientists interested in environmental change could learn from this. Fairhead and Leach (Chapter 10) show how our understanding of rain forests in West Africa has been influenced by a view of history that attributes all change to outsiders. By taking a longer perspective on people–vegetation interactions they clearly show the relationship to be dynamic and in a constant state of flux. Fears over the destruction of tropical rain forests is perhaps the most obvious example where a Eurocentric perspective has led to some patently ahistorical explanations; the existence of 'lost' forest cities, such as Angkor Wat in Cambodia, should have led more people to challenge the popular media image of a primeval forest. In Asia, Africa, and the Americas, histories of people–environment interactions tend to stress a great change at the time of the arrival of European colonists but to view the previous relationship as static.

In contrast to Fairhead and Leach (Chapter 10), who are primarily interested in the way changes in society have effected the biological world, Grove (Chapter 3) shows how people have responded to, or failed to respond to, climatic changes (the physical world) over a century time-scale (and she does so for both European and pre-colonial American societies). Her analysis shows how different pre-industrial societies were able to adapt their agricultural techniques in order to respond to climate change; the chapter also shows how only some people were able to benefit from these new techniques and that many lost out. All the contributors to this book recognise that it is impossible for us to stop environmental change. The obvious solution (made, for example by Wallis in Chapter 5) is therefore to suggest that we need to develop ways of adapting to and managing that change. But, as Grove (Chapter 3) shows, it is important to remember that not everybody will lose or benefit equally. New technology, such as remote sensing, has made predictions about climate change over a very short time-scale (weather) very much better but while it may be fairly easy to get timely information to a Mid-Western wheat farmer it is very difficult to get timely information to an Indian rice farmer. Over the century time-scale the balance of world resources and power will undoubtedly change, but (as W. Stanley Jevons was never able to know, but we living 130 years later can testify) predictions of what economic and political changes will occur are usually hopelessly wrong.

THE LATE QUATERNARY TIME-SCALE

Late Quaternary time-scales are most prevalent in models of explanation in the natural and physical sciences; physical geographers are very comfortable thinking about environmental change over these time-scales (see Roberts, Chapter 2). Despite the large literature on environmental change over these time-scales, long-term environmental change is rarely considered by people thinking about contemporary environmental change. One of the consequences of this is that the impact of humans on the environment is emphasised at the expense of the environment's own instability. There is, perhaps, a tendency in popular environmentalism to see studies of environmental change over these longer time-scales as purely an academic exercise, but the case study of dambo hydrology examined by Roberts (Chapter 2) shows the practical importance of considering the development of environmental features over longer time-scales.

At such time-scales as this we can sometimes think we can see patterns and correlations of great significance, and yet of course it is harder to prove or disprove any hypothesis. To many people it is not just chance that *Homo sapiens* has evolved in the Quaternary, because during this period the rapid and successive major fluctuations in climate and vegetation will have favoured the survival of an adaptive ape with great mental powers – although why the mutation which caused the cerebellum to expand so rapidly happened then and not at some other time is still unexplained. Whatever the truth of that, it is a simple fact that as we move through the late Quaternary and its rapid environmental change, so too we see the greatest changes in the application of technology by human beings and the greatest changes in their social organisation – from hunting and gathering in the early Neolithic period, to settled agriculture, metalworking, and city building, and the maintaining of written records. All of this means simultaneous change in the conscious world, and the world of artefacts, much of it directed to environmental manipulation. Although many civilisations have flourished and then faded, humankind in general has progressively flourished, as has the power of human thought to understand the changes we are enmeshed in. Such a view would give credence to the optimists who maintain that even if now the rate of environmental change again becomes stressful, and even if in consequences there are major demographic changes on the planet and major changes in political form, some societies will have the mental and social capacity to manage and to adapt and flourish.

TIME-SCALES AND MODELS IN THE PHYSICAL, NATURAL AND SOCIAL SCIENCES

'The environment' is a subject of enquiry in a vast range of disciplines, but there is a recognition of the need for an interdisciplinary approach that combines the physical, natural, and social sciences. In this final section we will examine some of the time-scale implications of the need to incorporate the very different models of explanation in the different sciences; we start with a re-examination of the international political debate over climate change.

Debates about global warming tend to be dominated by North–South disagreements; here, for the purposes of clarity, we have crudely characterised the two sides of the debate as the Northern and Southern perspectives and ignored the various nuances of the debate. Among members and allies of the Northern environmental movement there seems to be a perception that this disagreement is largely because Southern governments are even less able to take into account longer (generational) time-scales than Northern governments owing to the more pressing day-to-day development-related demands over the political time-scale ('now time'). The implication of this could be characterised in a time-scale context as the Northern environmental movements are considering a generational time-scale and Southern governments a 'now' time-scale. There is, however, another way of looking at the issue.

Using Chapman's analysis (in Chapter 11) the two different positions could be explained by considering whether the understanding starts from models of the physical/chemical world, specifically global circulation models (GCMs), and then moves towards models of human society (the world of conscious human thought) or vice versa. The North position starts with GCMs and then says that, given this undeniable 'truth', society, especially society in the South, must change to take this into account. The 'South' position, on the other hand, starts from models of society, especially models to explain global inequalities, and then relegates the environmental impact to a secondary position.

Models of understanding in the physical sciences tend to operate in the same manner at any point in time; that is, they are independent of history, and are often based on systems that can be run in either direction. Models in the sphere of social sciences (with the exception of economics), on the other hand, tend to be historical; they make sense only at a specific time and they can be run only forwards. The Northern perspective on climate change starts with models that run from 'now' (as we have shown above, this is a very problematic concept); the explanations then try to add on models of society.

The models of society are by necessity simple and not firmly rooted in history; they tend to be simply calls for the South to consider some sort of global responsibility.

The Southern position starts from models of society, in particular models that explain the uneven global consumption of resources. These are models that tend to be historical in nature; they explain the present situation in terms of past events. Many of the models favoured in the South trace the root of development and underdevelopment back into the colonial era; most of these run over a generational time-scale, but with a century time-scale backdrop of explanation. When models of the physical world are then considered from this perspective they tend to be either dismissed – they are after all the products of a Northern science that plays an integral role in perpetuating an unequal social world – or incorporated as yet another way in which the North, through its wasteful society, will adversely affect the South.

The Northern perspective is driven by the generational time-scale implicit in most GCMs; but it is an explanation that does not incorporate either models of the physical world that explain longer (century and late Quaternary) time-scale changes or models of society (over generational and century time-scales) that explain the global distribution of resources in historical terms. The Southern perspective, on the other hand, is driven by the generational (and sometimes century) time-scales implicit in models of society; but does not incorporate models of the physical world at either the (future-oriented) generational time-scale (GCMs) or the (past-oriented) century to late Quaternary time-scale. The North having got a fairly comfortable present, is not prepared to discount the future. The South, having a very uncomfortable present, is, by contrast, impelled to discount the future far more.

The nature of formal education systems means that even people heavily committed to interdisciplinary approaches to understanding environmental change start from within one discipline and then try to add on other perspectives. The different way in which time is conceptualised in different models of the world (see Table 11.1) makes this a very difficult task. What seems to be important is not just the scale of time that is considered but also how time itself is conceptualised. Explanations of people–environment relations that start from models of society take into account passing time only in relation to people (and sometimes only certain people); there is an expectation that the physical world remains the same and the biological world changes only very slowly. On the other hand, explanations of people–environment relations that start from models of the environment conceptualise passing time in relation to the physical world but tend to think about society in ahistorical terms. Perhaps people interested in time and environmental change need to ask not just 'how long?' but also 'how is time conceptualised?'.

NOTES
1   Dawson (1992) defines the late Quaternary as the past 130,000 years.
2   By this we are in no way stressing a Eurocentric way of thinking about the development of knowledge; the process of exploration of other continents is, of course, something that happened long before the development of European expansionism over the past five hundred years or so.
3   Cooke (1992) has suggested that both of these topics are areas needing increased research interest – and identifies environmental change over the century time-scale as an important research lacuna.
4   There is, of course, debate over how 'Darwinian' the social Darwinism of scientific racism is; it is used here as a convenient shorthand. The expanding consciousness of time that resulted from Hutton's explanation of the history of the earth (see Chapter 1) was important in the development of race science: 'The idea of the Great Chain of Being, a hierarchy of living forms with man at the top of the material order, could now be conceived chronologically, with increasingly sophisticated forms superseding each other through the eons' (Kohn 1995, p. 29).
5   Probably inevitably, Fernández-Armesto's book does not quite live up to its billing. One of its more glaring flaws is totally ignoring the effects of environmental change, with the exception of one fleeting reference to the Little Ice Age in Europe.

REFERENCES

Behnke, R. H.. Scoones, I., and Kerven, C. 1993. *Range Ecology at Disequlibrium: New Models of Natural Variability and Pastoral Adaptation in African Savannas*. London: Overseas Development Institute.

Cooke, R. 1992 Common ground, shared inheritance: research imperatives in environmental geography. *Transactions of the Institute of British Geographers*, n.s. 17: 131–51.

Crosby, A. W. 1986. *Ecological Imperialism: The Biological Expansion of Europe, 900–1400*. Cambridge: Cambridge University Press.

Dawson, A. G. 1992. *Ice-Age Earth: Late Quaternary Geology and Climate*. London: Routledge.

Fernández-Armesto, F. 1995. *Millennium: A History of our Last Thousand Years*. London: Bantam.

Kohn, M. 1995. *The Race Gallery: The Return of Racial Science*. London: Jonathan Cape.

World Commission on Environment and Development 1987. *Our Common Future*. Oxford: Oxford University Press.

INDEX

Hall, T. D. 153, 154, 162
Hamilton, Sir William Rowan 198, 216
Harvey, A. C. 90, 101
Hasselmann, K. H. 172, 177, 189
Hindu cosmology 5, 6, 7–8, 223, 225
history/historical time 3, 19–20, 26,
    150–3, 160–1, 222–5, 264; *see also*
    forest history
Hoffman, M. T. 150, 160–1, 162, 164, 166
Holocene 26, 40–54; climate change 30,
    31–2, 35–6; fluctuations 26, 40–4;
    glacial history 44–54
homogeneity of time series 91, 92–6,
    99–100
Howarth, R. B. 113, 118
humanity *see* mankind and earth
hunting and fishing 55–8, 151
Huntley, B. 116, 157
Hutton, James 10–11, 12, 268

Ice Ages 39–40, 122; cycles 17–18, 28; *see
    also* Holocene; Pleistocene
ice cores, analysis of 41, 51, 55, 91, 116
impact assessment 153–8
India/South Asia 2, 24, 158, 218–55, 258,
    261; desert 28, 29, 30; environment
    225–31, 234–7; history 222–5; media
    and public opinion 222, 225, 236,
    237–53; social change 231–3;
    time-scales 40, 42, 46, 263, 266, 267;
    units of time 4
individual, importance of 19, 201; *see
    also* lifetime
industry *see* development; economic
    action
instrumental record 16, 23, 88–107, 259,
    263; *see also* CET
Intergovernmental Agreement on
    Environment (IGAE) 126, 138
Intergovernmental Panel on Climate
    Change *see* IPCC
intermediate time-scales 26
international action 121–4, 221
international awareness in India lacking
    241, 242, 252, 253, 254
International Development Decade 20
intuitive human experience of time 4–6
invention and innovation 208–11, 216
IPCC (Intergovernmental Panel on
    Climate Change) 81, 109–10, 111,
    116, 259; greenhouse policy in

Australia 122, 123–4, 126, 134, 136–7,
    138, 143
irrigation 33–4, 152, 220, 227, 230, 231
islands of forest in savanna 170, 178–84,
    190
isotopes *see under* radioactivity

Jenkins, G. M. 101, 103
Jevons, W. Stanley 257, 258, 265
Jones, P. D. 51, 53–4, 88
Judaic–Christian cosmology 5, 6–7, 8–9,
    10, 12

Karshenas, M. 234, 236–7, 254
Keay, R. W. J. 179, 180, 182
Kelly, M. 112, 116
Kelly, Ros 126, 127, 133
Kershaw, I. 65, 66, 69, 82

lakes: levels, salinity and sedimentation
    26, 27, 29, 30, 32, 187
language and communication 6; India
    237–8, 252, 254
Latin America 22, 29, 158, 224, 264;
    century time-scale 40, 42, 46;
    population 231, 232
Laurmann, J. A. 111, 113, 118
Leach, Melissa: on forest 23, 169–95, 262
lifetime time-scale 4–5, 21, 147; *see also*
    southern Africa
Lind, R. C. 113, 118, 216
Little Ice Age 32, 40, 82, 91, 92; in Crete
    71–80, 81
livestock 34–5; Medieval Warm Epoch
    54–5, 57, 66, 71, 82; North America
    152, 158, 160, 163, 167; *see also*
    southern Africa
local: awareness in India 241, 244, 253;
    inhabitants and forest history 170,
    178–84
long-term 5; environmental stability and
    instability in tropics and subtropics
    23, 25–38, 149, 265 (*see also* dambos;
    Milankovich); instrumental record
    91; need for 141, 142–3; *see also*
    geology
Lorenz, K. 16, 17, 259
Lorius, C. 28, 39
Lovelock, J. 13, 219
Lucas, N. J. D. 110–11, 114

MA (moving average) models 101–2, 103

Milton Keynes UK
Ingram Content Group UK Ltd.
UKHW040013071024
449327UK00011B/213